Watershed Water Environment and Hydrology under the Influence of Anthropogenic and Natural Processes

Watershed Water Environment and Hydrology under the Influence of Anthropogenic and Natural Processes

Editors

Guilin Han
Zhifang Xu

MDPI • Basel • Beijing • Wuhan • Barcelona • Belgrade • Manchester • Tokyo • Cluj • Tianjin

Editors
Guilin Han
China University of
Geosciences (Beijing)
China

Zhifang Xu
Institute of Geology and
Geophysics, Chinese
Academy of Sciences
China

Editorial Office
MDPI
St. Alban-Anlage 66
4052 Basel, Switzerland

This is a reprint of articles from the Special Issue published online in the open access journal *Water* (ISSN 2073-4441) (available at: https://www.mdpi.com/journal/water/special_issues/watershed_water_environment).

For citation purposes, cite each article independently as indicated on the article page online and as indicated below:

LastName, A.A.; LastName, B.B.; LastName, C.C. Article Title. *Journal Name* **Year**, *Volume Number*, Page Range.

ISBN 978-3-0365-4059-7 (Hbk)
ISBN 978-3-0365-4060-3 (PDF)

Contents

About the Editors

Guilin Han (Prof.) is currently a professor at China University of Geosciences (Beijing). She received her Ph. D. in environmental geochemistry at the Institute of Geochemistry, Chinese Academy of Sciences in Guiyang. She studies the surficial environment geochemistry in China. Her major contributions include developing methods for purifying K, K-Ca-Sr from a geological matrix for precise isotope analysis, measuring the Ca and Sr isotopic composition of rainwater in different ecosystems in China, and watershed weathering and global C cycles based on the Sr, C isotope. She has published many papers during her career in journals such as *Earth and Planetary Science Letters, Geochimica et Cosmochimica Acta, Water Research, Chemical Geology, Atmospheric Environment, Sciences of Total Environment, Journal of Hydrology, Journal of Analytical Atomic Spectrometry*, and *Agriculture Ecosystems and Environment*, among others.

Zhifang Xu (Prof.) is currently a professor at the Institute of Geology and Geophysics, CAS. He received his Ph. D. in Geology at the Institute of Geology and Geophysics, CAS in Beijing. He studies the surficial environment geochemistry in China. His major contributions to the field include research on continental weathering and carbon cycling, a watershed biogenic elements cycle, and sources and environmental effects of precipitation. He has published numerous papers in journals such as *Geochimica et Cosmochimica Acta, Biogeosciences, Chemical Geology, Atmospheric Environment, Sciences of Total Environment, Journal of Hydrology*, and *Geochemistry Geophysics Geosystems*, among others.

Editorial

Watershed Water Environment and Hydrology under the Influence of Anthropogenic and Natural Processes

Guilin Han [1,*] **and Zhifang Xu** [2]

1 Institute of Earth Sciences, China University of Geosciences (Beijing), Beijing 100083, China
2 Institute of Geology and Geophysics, Chinese Academy of Sciences, Beijing 100029, China; zfxu@mail.iggcas.ac.cn
* Correspondence: hanguilin@cugb.edu.cn

Citation: Han, G.; Xu, Z. Watershed Water Environment and Hydrology under the Influence of Anthropogenic and Natural Processes. *Water* 2022, 14, 1059. https://doi.org/10.3390/w14071059

Received: 21 March 2022
Accepted: 24 March 2022
Published: 28 March 2022

Publisher's Note: MDPI stays neutral with regard to jurisdictional claims in published maps and institutional affiliations.

1. Introduction to Watershed Water Environment and Hydrology

Water resources imbalance of requirement and distribution has become one of the most vital limiting factors for regional and global sustainable development [1,2]. Under the global environmental change, water quality, including pollution risk, identification, and controlling factors, becomes the key to sustainable development and high-efficiency use of water resources. Watersheds are closely related to social and economic development around the world, as the most critical water resource unit in surface systems. However, the global water eco-environment poses many challenges for protection and management at different watershed scales. Under the influence of anthropogenic and natural processes, nutrients and other pollutants have difficulty being assessed accurately with respect to their transformation and migration at different watershed scales. By studying the biogeochemical cycle of substances and sources of pollutants in the water environment (e.g., rivers, reservoirs, and subterranean rivers) at the watershed scale, combined with hydrology methods, the mechanism of ecological environment changes at watershed scale can be explored under the influence of both anthropogenic and natural processes [3,4].

Generally, human activities (such as agricultural production, urban sewage, industrial discharge, and mining in the watershed) are the main sources of pollutants in the water environment, and natural processes (mainly rock weathering) are also important factors controlling the water chemistry of watersheds [5–7]. Meanwhile, besides natural hydrological processes, reservoir and dam construction (water conservancy projects) and land-use change are also important factors affecting material migration and transformation in the watershed water environment. In this Special Issue, we aim to promote paper publications that deal with watershed water environments and hydrology under the influence of anthropogenic and natural processes, mainly focusing on the quality and contamination of water bodies and their influencing factors. This Special Issue shares innovative/new ideas on the watershed water environment from different perspectives across the field.

The aims of this Special Issue are to (1) distinguish the evolution of watershed water ecological and environmental quality impacted by both anthropogenic and natural processes; (2) clarify the biogeochemical cycling of elements or pollutants driven by human activities and hydrological factors at watershed scale; (3) identify and quantify the sources of pollutants in watershed water environments; (4) assess the ecological risk and human health risk of pollutants in the water environment at different watershed scales.

A total of eight peer-reviewed articles were collected in this Special Issue. Six papers come from China, and the other two were from the United States of America and Russia, respectively. Overall, these papers in this Special Issue point out several perspectives of watershed water environment and are of great significance for realizing high-efficiency water environmental management and sustainable use of water resources. The research subjects involved different rivers and river-reservoir systems in China, the USA, Russia, and Sri Lanka, such as the Yangtze River Watershed and Wujiang River Basin in China, Triangle Area in the USA, and Nizhnekamskoe Reservoir Watershed in Russia.

2. Overview of This Special Issue

The topics of collected papers in this Special Issue are widespread and they can be divided into three parts as follows: (1) hydrochemistry-based watershed weathering processes and their environmental implications, (2) trace elements in watershed water environment and their risks, and (3) nutrients cycle in river-reservoir systems.

2.1. Hydrochemistry-Based Watershed Weathering Processes and Its Environmental Implications

Chemical weathering adjusts atmospheric CO_2 balance and the habitability of Earth-surface on a long-term scale. As the integration of solid and dissolved weathering products, river geochemistry can reflect the weathering processes and fluxes at the basin scale. Moreover, the river water hydrochemical compositions are not only the reflection of natural processes (mainly weathering processes), but also the carrier of environmental information of human activities.

Two papers deal with the water chemistry of river watersheds, focusing on the hydrochemical behavior under different basin backgrounds and geological conditions.

Ge et al. [8] reported the hydrochemistry and sulfur isotope (δ^{34}S-SO_4^{2-}) compositions of a representative carbonate rock area to identify the potential origins of fluvial solutes (mainly weathering products), the human disturbance, and river water quality. The findings presented that K^+, Mg^{2+}, F^-, HCO_3^- mainly reflected the rock weathering inputs, and atmospheric deposition was the contributor of Na^+ and Cl^-, while SO_4^- and NO_3^- were defined as the anthropogenic inputs. The H_2SO_4-involved processes were significantly facilitated by weathering processes. The sulfur isotope-based discussion further revealed that human emission controlled the fluvial SO_4^{2-} relative to sulfide oxidation, while the atmospheric impact was negligible.

Wang et al. [9] applied the Germanium/Silicon (Ge/Si) ratio to trace the influence of hydrothermal input and chemical weathering on Tibetan Plateau-originated river water (Yarlung Tsangpo River). Based on that Ge/Si ratio, this paper reflected the primary mineral dissolution and secondary clay formation. The main results highlight that the hydrothermal water contribution notably impacts the Ge/Si ratios of river water, particularly in the upper-mid reaches. About 11–88% of Ge was lost during the transported processes from hydrothermal water to river system. The contribution of hydrothermal sources should be considered if the Ge/Si ratio was used to trace silicate weathering in Tibetan Plateau rivers.

2.2. Trace Elements in Watershed Water Environment and Their Risks

The trace elements, including heavy metals (HMs) and rare earth elements (REEs) in the river system, are important and gain more concerns due to their potential toxicity on aquatic organisms and humans. Four papers in this issue determine the trace elements geochemistry in different watersheds to explore the natural and anthropogenic inputs of trace elements, and their potential health risks.

Zeng et al. [10] investigated the distribution, status, and sources of typical HMs, and further assessed the water quality and HMs-related risks in the upper Three Gorges Reservoir (TGR) of Yangtze River. The detectable HMs were 1.4~8.1 times higher than the source area of the Yangtze River, indicating potential anthropogenic inputs. V, Cu, As, and Pb concentrations along the main channel were decreased. Principal component analysis-based sources identification reflected that V, Ni, As, and Mo were the main contribution of human inputs, Cu and Pb were mainly from mixed sources of human emission and natural process, while the Zn and Cd were controlled by natural sources. Water quality assessment revealed a good water quality for the drinking purpose with limited exposure risk.

The article by Zabrecky et al. [11] is a good application of REEs (mainly Gadolinium, Gd) tracing the anthropogenic contributions on river water, in particular, the influences of surrounding wastewater treatment plants (WWTPs). The Gd anomalies were investigated in North Carolina's Triangle Area. The quantified assessment of Gd in wastewater samples suggested that 98.1% to 99.8% Gd is an anthropogenic contribution, while the anthropogenic Gd contribution of upstream and downstream samples was estimated as an average increase of 45.3%.

Motovilov et al. [12] simulated the heavy metals (Cu, Zn, and Mn) cycling of the heavily polluted Nizhnekamskoe reservoir basin in Russia using the semi-distributed physically based ECOMAG-HM model. The spatial and temporal patterns of these HMs were also identified. The findings reflected that the riverine pollution is formed mainly by the metals diffuse wash-off into rivers from the soil-ground layer. The delta-change climatic scenario-based model highlighted that water quality characteristics should not be significantly changed up to 2050.

Xu et al. [13] described the spatial characteristics of groundwater geochemistry in Sri Lanka to link the water quality and the potential contributor to chronic kidney disease of unknown etiology (CKDu). The groundwater geochemistry exhibited significant spatial heterogeneity in Sri Lanka, and Cd, Pb, Cu, and Cr concentrations were within the limitation of World Health Organization guideline values. In contrast, the As and Al concentrations of some samples were higher than the limited values. Although the water quality data cannot explore if water quality is associated with the CKDu occurrence, more effective research should be conducted to confirm the synergistic effect of different chemical constituents on CKDu.

2.3. Nutrients Cycle in River-Reservoir Systems

The anthropogenic pollution of nitrogen and phosphorus in river watershed scales has become a global awareness due to their negative influence on aquatic quality. Given that the enrichment of N and P in the river-reservoir system has seriously affected the health of river water, and the knowledge of the main influencing factors controlling the N and P cycle is vitally significant. In this Special Issue, two papers aimed at the transportation and transformation of N and P in the river-reservoir system.

Hou et al. [14] performed their study on the nitrogen species and isotopes in the sediment of a deep reservoir (artificial reservoir of Wujiang River basin) in Southwest China to better understand nitrogen transformation under the condition of thermal stratification, in particular, the sediment–water interface (SWI) nitrogen cycle. This paper highlighted that the changes in dissolved oxygen primarily controlled the nitrogen cycling processes, and furthermore, nitrification, denitrification, mineralization, and diffusion of nitrogen species were largely varied with the presence of the oxygen.

The paper by Zhou et al. [15] investigated five sediment cores from the shallow YuQiao Reservoir in northern China to clarify the role of phosphorus (P) release from sediment by determining the characteristics and P fractions. The sediments presented a P sorption capacity of 7–10 times that of soil. The isotherm adsorption experiments identified the sediment contributes with a positive flux of P to the overlying water. The dredging of 30 cm surface sediments was the effective pathway to decline the soluble reactive phosphate.

3. Conclusions

Watershed water environments are complex earth–surface systems, as they integrate the influences of both natural processes and anthropogenic processes, which are also amongst the most significant freshwater resources, supporting the development of human societies over the world. This Special Issue collected eight peer-reviewed articles to clarify the watershed water environment and hydrology under the influence of anthropogenic and natural processes and further support the water eco-environmental protection and management on different watershed scales. These works cover several river watersheds and/or river-reservoir systems in China, the USA, Russia, and Sri Lanka. From the perspective of hydrochemistry-based weathering processes, trace elements and their risks, and nutrients cycle in different watersheds, the authors provide several benefits innovative/new ideas on watershed water environment and further help manage and balance the water resources.

Author Contributions: G.H. and Z.X. contributed to this manuscript. All authors have read and agreed to the published version of the manuscript.

Funding: This editorial was funded by the National Natural Science Foundation of China (No. 41661144029, No. 41325010 and No. 41730857).

Conflicts of Interest: The authors declare no conflict of interest.

References

1. Zeng, J.; Han, G.; Yang, K. Assessment and sources of heavy metals in suspended particulate matter in a tropical catchment, northeast Thailand. *J. Clean. Prod.* **2020**, *265*, 121898. [CrossRef]
2. Liu, J.; Han, G. Tracing Riverine Particulate Black Carbon Sources in Xijiang River Basin: Insight from Stable Isotopic Composition and Bayesian Mixing Model. *Water Res.* **2021**, *194*, 116932. [CrossRef]
3. Zeng, J.; Han, G.; Zhang, S.; Liang, B.; Qu, R.; Liu, M.; Liu, J. Potentially toxic elements in cascade dams-influenced river originated from Tibetan Plateau. *Environ. Res.* **2022**, *208*, 112716. [CrossRef]
4. Liu, X.; Han, G.; Zeng, J.; Liu, J.; Li, X.; Boeckx, P. The effects of clean energy production and urbanization on sources and transformation processes of nitrate in a subtropical river system: Insights from the dual isotopes of nitrate and Bayesian model. *J. Clean. Prod.* **2021**, *325*, 129317. [CrossRef]
5. Liu, X.-L.; Han, G.; Zeng, J.; Liu, M.; Li, X.-Q.; Boeckx, P. Identifying the sources of nitrate contamination using a combined dual isotope, chemical and Bayesian model approach in a tropical agricultural river: Case study in the Mun River, Thailand. *Sci. Total Environ.* **2021**, *760*, 143938. [CrossRef] [PubMed]
6. Liu, J.; Han, G. Tracing riverine sulfate source in an agricultural watershed: Constraints from stable isotopes. *Environ. Pollut.* **2021**, *288*, 117740. [CrossRef] [PubMed]
7. Li, X.; Han, G.; Liu, M.; Liu, J.; Zhang, Q.; Qu, R. Potassium and its isotope behaviour during chemical weathering in a tropical catchment affected by evaporite dissolution. *Geochim. Cosmochim. Acta* **2022**, *316*, 105–121. [CrossRef]
8. Ge, X.; Wu, Q.; Wang, Z.; Gao, S.; Wang, T. Sulfur Isotope and Stoichiometry–Based Source Identification of Major Ions and Risk Assessment in Chishui River Basin, Southwest China. *Water* **2021**, *13*, 1231. [CrossRef]
9. Wang, Y.; Zhao, T.; Xu, Z.; Sun, H.; Zhang, J. Ge/Si Ratio of River Water in the Yarlung Tsangpo: Implications for Hydrothermal Input and Chemical Weathering. *Water* **2022**, *14*, 181. [CrossRef]
10. Zeng, J.; Han, G.; Hu, M.; Wang, Y.; Liu, J.; Zhang, S.; Wang, D. Geochemistry of Dissolved Heavy Metals in Upper Reaches of the Three Gorges Reservoir of Yangtze River Watershed during the Flood Season. *Water* **2021**, *13*, 2078. [CrossRef]
11. Zabrecky, J.M.; Liu, X.-M.; Wu, Q.; Cao, C. Evidence of Anthropogenic Gadolinium in Triangle Area Waters, North Carolina, USA. *Water* **2021**, *13*, 1895. [CrossRef]
12. Motovilov, Y.; Fashchevskaya, T. Modeling Management and Climate Change Impacts on Water Pollution by Heavy Metals in the Nizhnekamskoe Reservoir Watershed. *Water* **2021**, *13*, 3214. [CrossRef]
13. Xu, S.; Li, S.-L.; Yue, F.; Udeshani, C.; Chandrajith, R. Natural and Anthropogenic Controls of Groundwater Quality in Sri Lanka: Implications for Chronic Kidney Disease of Unknown Etiology (CKDu). *Water* **2021**, *13*, 2724. [CrossRef]
14. Hou, Y.; Liu, X.; Chen, S.; Ren, J.; Bai, L.; Li, J.; Gu, Y.; Wei, L. Effects of Seasonal Thermal Stratification on Nitrogen Transformation and Diffusion at the Sediment-Water Interface in a Deep Canyon Artificial Reservoir of Wujiang River Basin. *Water* **2021**, *13*, 3194. [CrossRef]
15. Zhou, B.; Fu, X.; Wu, B.; He, J.; Vogt, R.D.; Yu, D.; Yue, F.; Chai, M. Phosphorus Release from Sediments in a Raw Water Reservoir with Reduced Allochthonous Input. *Water* **2021**, *13*, 1983. [CrossRef]

 *water*

Article

Sulfur Isotope and Stoichiometry–Based Source Identification of Major Ions and Risk Assessment in Chishui River Basin, Southwest China

Xin Ge [1], Qixin Wu [1,*], Zhuhong Wang [2], Shilin Gao [1] and Tao Wang [1]

[1] Key Laboratory of Karst Geological Resources and Environment, College of Resources and Environmental Engineering, Guizhou University, Guiyang 550025, China; april_gexin@163.com (X.G.); gaoshilin110@163.com (S.G.); wangtao6090@126.com (T.W.)

[2] Key Laboratory of Environmental Pollution and Disease Monitoring of Ministry of Education, School of Public Health, Guizhou Medical University, Guiyang 550000, China; cindywzh@163.com

* Correspondence: qxwu@gzu.edu.cn

Abstract: Hydrochemistry and sulfur isotope ($\delta^{34}S$–SO_4^{2-}) of Chishui River watershed in Southwest China were measured to identify the sources of riverine solutes, the potential impact of human activities, water quality, and health risk. The main findings indicated that the HCO_3^- (2.22 mmol/L) and Ca^{2+} (1.54 mmol/L) were the major ions, with the cation order of Ca^{2+} (71 ± 6%) > Mg^{2+} (21 ± 6%) > $Na^+ + K^+$ (8 ± 3%) and the anion sequence of HCO_3^- (55 ± 9%) > SO_4^{2-} (41 ± 9%) > Cl^- (4 ± 3%). The riverine $\delta^{34}S$–SO_4^{2-} values fluctuated from −7.79‰ to +22.13‰ (average +4.68‰). Overall, the water samples from Chishui River presented a hydrochemical type of Calcium–Bicarbonate. The stoichiometry and PCA analysis extracted three PCs that explained 79.67% of the total variances. PC 1 with significantly positive loadings of K^+, Mg^{2+}, F^-, HCO_3^- and relatively strong loading of Ca^{2+} revealed the natural sources of rock weathering inputs (mainly carbonate). PC 2 (Na^+ and Cl^-) was primarily explained as atmospheric contribution, while the human inputs were assuaged by landscape setting and river water mixing processes. The strongest loadings of SO_4^{2-} and NO_3^- were found in PC 3, which could be defined as the anthropogenic inputs. The H_2SO_4–involved weathering processes significantly impacted (facilitated weathering) the concentrations of riverine total ions. Sulfur isotope compositions further indicated that riverine SO_4^{2-} were mainly controlled by anthropogenic inputs SO_4^{2-} compared to the sulfide oxidation derived SO_4^{2-}, and the atmospheric contribution was very limited. The results of risk and water quality assessment demonstrated that Chishui River water was desirable for irrigation and drinking purposes due to low hazard quotient values (<1, ignorable risk), but long–term monitoring is still worthy under the circumstances of global environmental change.

Keywords: water chemistry; sulfur isotope; ion source apportionment; water quality and risk assessment; Chishui River watershed

Citation: Ge, X.; Wu, Q.; Wang, Z.; Gao, S.; Wang, T. Sulfur Isotope and Stoichiometry–Based Source Identification of Major Ions and Risk Assessment in Chishui River Basin, Southwest China. *Water* **2021**, *13*, 1231. https://doi.org/10.3390/w13091231

Academic Editor: Frédéric Huneau

Received: 22 March 2021
Accepted: 25 April 2021
Published: 28 April 2021

Publisher's Note: MDPI stays neutral with regard to jurisdictional claims in published maps and institutional affiliations.

1. Introduction

Watershed–scale hydrochemical evolution and water environmental quality, the most important hot point for river water resource research, is the basis of effective hydrospheric environment planning and sustainable usage of freshwater resources, in particular, under the background of the imbalance between unevenly distributed water resources and water requirements. Both anthropogenic and natural processes can affect hydrochemistry and the major ion concentration level [1,2]. With the accelerated improvement of economy, the increased anthropogenic activities (e.g., industrial/domestic wastewater, agricultural emissions) have rapidly enhanced the riverine pollutants [3,4], which will further result in a series of environmental issues and risks [5,6], such as the destruction of soil aggregate structure by polluted river water irrigation, and the health risks via drinking water [7–9].

The major ions are the most important part of river dissolved loads. Exploring the origins of these ions could greatly benefit for the understanding of watershed–scale geochemical dynamics (e.g., weathering processes) and distribution/transformation regulation of pollutants. Moreover, the ion source identification is also benefit for the environmental supervision of the government [10]. Generally, in addition to regularly direct monitoring of cation and anion concentrations, statistic approaches, ion ratios, and isotopic methods are widely applied in source identification [11]. Statistical methods, such as principal component analysis and correlation analysis could explore the common sources of riverine solutes due to similar physicochemical characteristics and potential co–origins of some ionic species [12]. The potential dilution effect can be avoided by the fluvial ion ratios, reflecting the mixing processes of the sources [13]. The isotopic methods, such as sulfur (S) isotope of riverine SO_4^{2-}, could distinguish different sources via specific isotope compositions of these end–members, based on the inheritance of S isotopes of riverine sulfate from both natural and anthropogenic sources [2,14].

In the karst area of southwest China, the typical carbonate geomorphology is developed, such as the sinkhole, depression, cockpit, and cone, which further result in the barren and thin soil [15,16]. Thus, the karstic ecosystem, in particular river system is extremely sensitive and vulnerable due to the unique geological conditions and strong karstification [17,18]. In order to obtain more knowledge of karst fluvial hydrochemistry, this study presents the detailed investigation of river water stoichiometry and S isotope of Chishui River (a tributary of upper Yangtze River) watershed in southwest China karst region. The main aims are to:

(i) Carify the ionic compositions and S isotope compositions of river water,
(ii) identify the source of major ions, and
(iii) explore the water quality as well as the potential irrigation and health risks.

2. Materials and Methods

2.1. Study Region

The Chishui River is one of the first–level tributaries in upper Yangtze River that has not been dammed in the main channel. With a total length of 445 km for main stream, the Chishui River begins in the Zhenxiong County, Yunnan Province and flows from southwest to northeast through Yunnan–Guizhou Plateau and Sichuan Basin in southwestern China, and finally enters into the Yangtze River in Hejiang County, Sichuan Province (Figure 1) [7]. In Chishui River watershed (CRW, 27°15′~28°50′ N, 104°44′~107°1′ E), there is a wide catchment area (~1.89 × 10^4 km^2) with various tributaries influenced by different degrees of human activities. As shown in Figure 1a, the lithology exposed in CRW is mainly composed of carbonate sedimentary rocks (~44.6%), siliciclastic sedimentary rocks (~24.5%), mixed sedimentary rocks (~30.1%), and basic volcanic rocks (~0.7%). In more detail, the strata distributed in upper stream are mainly dolomite of Dengying Formation of Ediacaran with the minerals resources of phosphorite, barite, and fluorite. The middle reaches are mainly composed of limestone and dolomite of Cambrian, Ordovician, Silurian, Permian, Triassic, and Jurassic, followed by mudstone, sand shale and coal–bearing rock group. The middle reaches also developed the minerals resources of coal, pyrite (iron sulfide), gypsum, and barite with the same geological age. The lower stream mainly distributed Jurassic–Cretaceous siltstone and mudstone with sporadic oil shale. The CRW covers various land use and a large elevation range (205 to 2237 m). The land use of CRW mainly includes water area, urban area, grass land, unused land, cropland, and forest land (Figure 1b). Forest land (~73.4%) and cropland (~19.6%) are the dominant land uses in CRW. The CRW is affected by the sub–tropical monsoon climate, with the average annual air temperature of 11~13 °C. The annual rainfall ranges from 800 to 1200 mm. The agriculture is distributed in upper–middle stream with a vulnerable eco–environment, while the industries are mainly located in the middle stream. The water quality of CWR is of great importance due the significant supporting role of water resources in the watershed.

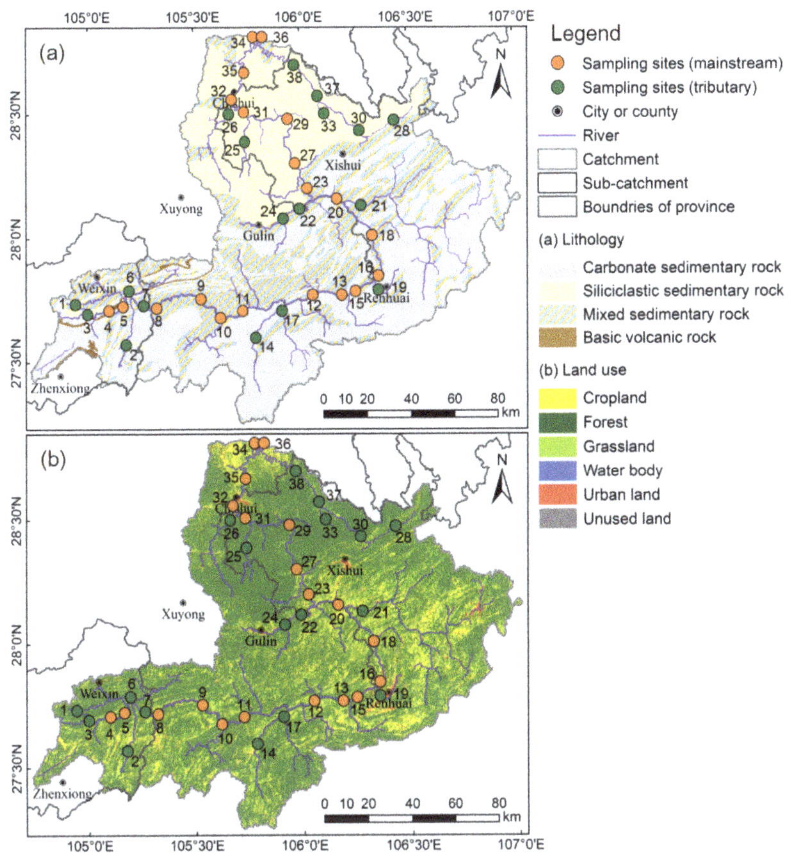

Figure 1. The background of sampling sites: (**a**) the lithology distribution of Chishui River watershed; (**b**) the land use of Chishui River watershed.

2.2. Sample Collection and Analyses

A systematic watershed survey was conducted from November to December in 2012 (dry season was selected to avoid the dilution effect and biologic effect as much as possible). According to the natural features (landuse and lithology) and the population distribution, 38 sample sites were chosen in the mainstream and tributaries of CRW (2 sites were chosen in Changjiang River, Figure 1). In total, 38 river water samples were obtained and further filtered (0.45–μm membrane, Millipore) and saved in the clean polyethylene sample bottles. Each sample was divided into three parts for the measurement of sulfur isotope and the major ions. The storage conditions of all samples were lightless and refrigerated (4 °C). The river water parameters (pH, T, DO, and EC) were measured in the field using the multi–parameter meter (Multi Line 3320, WTW, Munich, Bavaria, Germany). HCO_3^- concentrations were detected by HCl–titrated method. The Na^+, K^+, Mg^{2+}, Ca^{2+}, F^-, Cl^-, NO_3^-, and SO_4^{2-} concentrations were measured by ion chromatograph (ICS–1100, Thermo Fisher, Waltham, MA, USA) at the Key Laboratory of Karst Geological Resources and Environment, Ministry of Education, Guizhou University. The analysis was conducted with replicate samples and procedural blanks to maintain the accuracy of measurement. The measured result of replicate samples suggested acceptable repeatability for all ions (relative standard deviations were within ±5%), and the procedural blanks for all ions were generally lower than the detection limit.

The sulfur isotope of riverine SO_4^{2-} was detected according to previous studies [19]. Briefly, 10% $BaCl_2$ solution were added to water samples to convert the dissolved SO_4^{2-} into $BaSO_4$ precipitation, and the mixture was filtered by 0.45–μm membrane filters after 48 h. Then, the $BaSO_4$ precipitation on membranes was further calcined (800 °C, 40 min) to gain the $BaSO_4$ solid. The S isotope compositions (in δ notation relative to Vienna–Canyon Diablo Troilite, VCDT; Equation (1)) were detected by Elemental Analyzer–IsoPrime MS (IsoPrime, GV Instruments, Manchester, UK) at the Institute of Geochemistry, Chinese Academy of Sciences [20].

$$\delta^{34}S\ (\text{‰}) = (R_{\text{sample}}/R_{\text{VCDT}} - 1) \times 1000 \tag{1}$$

The standard reference materials (NBS 127) and internal laboratory standard reference materials (pre–calibrated Lab–SO1, Lab–SO4, Lab–Sigma) for S isotope were also detected to ensure the analysis accuracy, which were within the recommended value (2SD smaller than 0.02‰).

2.3. Assessment Method

Chishui River is the most important water source of both drinking and agricultural irrigation water within the watershed, which is necessary to assess the suitability for drinking and irrigation purpose. The riverine parameters (pH) and ion concentrations of Chishui River water were appraised by Chinese (GB 5749–2006) and WHO drinking water quality guidelines. For the irrigation water quality, the soil quality attributes can be influenced by the different salinity and alkalinity level of irrigation water, and further changed the yields from farmland. To gain an overall evaluation of irrigation water salinity and alkalinity hazard, the commonly–applied indicators such as sodium adsorption ratio (SAR), soluble sodium percentage (Na%), and residual sodium carbonate (RSC) were calculated based on the ion equivalent concentrations (meq/L) of river water as follows [11]:

$$SAR = Na^+ \times [2/(Mg^{2+} + Ca^{2+})]^{0.5} \tag{2}$$

$$Na\% = 100\% \times Na^+/(Na^+ + K^+ + Mg^{2+} + Ca^{2+}) \tag{3}$$

$$RSC = (HCO_3^- + CO_3^{2-}) - (Mg^{2+} + Ca^{2+}) \tag{4}$$

Regarding to the human health risk, the health threat could be occurred via ingestion and dermal exposure to river water with high–concentration ions [21]. In comparison, the health risk through ingestion intake is the most noteworthy. To assess this threat (non–carcinogenic health risk), the Hazard quotient (HQ) suggested by the U.S. Environmental Protection Agency was calculated for F^-, NO_3^-, and NH_4^+ [11]:

$$ADD_{\text{ingestion}} = C \times IR \times EF \times ED/(BW \times AT) \tag{5}$$

$$HQ = ADD/RfD \tag{6}$$

where $ADD_{\text{ingestion}}$, C, IR, EF, ED, BW, AT, and RfD represents the daily ingestion intake doses, ion concentrations (mg/L), daily ingestion rate (0.6 L/day for children, 1.0 L/day for adults), exposure frequency (365 days/year), exposure duration (12 and 25 years for children and adults), body weight (16 and 56 kg for children and adults), average time for non–carcinogenic effect (4380 and 10950 days for children and adults), reference dose of different ions (0.04, 1.6, and 0.97 ppm/day for F^-, NO_3^-, and NH_4^+), respectively. If the HQ of each ion (or total hazard quotient, HQ_t) is >1, the local people will be threatened by the corresponding ion–derived risk.

2.4. Software for Data Analyses

For the statistical analyses of riverine ion concentrations and other parameters, the principal component analysis (PCA, a common method for exploring the potential origins of ions) and Piper diagram analysis were carried out by SPSS 21.0. All the data–based figures were illustrated by Origin 2018.

3. Results and Discussion

3.1. Overview of Hydrochemical Compositions

The hydrochemical compositions and parameters of all river water samples are presented in Table S1, with the statistical results, including ranges, and mean values of these data summarized in Table 1. The pH values ranged from 7.63 to 8.87 (mean value 8.35), presenting the slightly alkaline river waters, further indicated that the inorganic carbon species was dominated by bicarbonate (HCO_3^-) based on the carbonate equilibrium, while CO_3^{2-} and dissolved CO_2 were negligible. The EC value averaged 403 µS/cm with a range of 128~738 µS/cm. The ionic balance was also applied for the river water samples, that is, the total cation concentration ($TZ^+ = Na^+ + K^+ + Mg^{2+} + Ca^{2+}$, in meq/L) and total anion concentration ($TZ^- = F^- + Cl^- + NO_3^- + SO_4^{2-} + HCO_3^-$, in meq/L) of river water, which presented well ion balances ($R^2 = 0.987$, $p < 0.01$). The normalized ionic charge balance ($[TZ^+ - TZ^-]/TZ^-$) is smaller than 11% (Table S1), suggesting the potential effect of the organic anionic species [22] like oxalate.

Table 1. Statistical result of hydrochemical compositions, parameters, $\delta^{34}S$ values, SAR, Na%, and RSC of Chishui River water.

	Unit	Min	Max	Mean	SD	Chinese Guideline [b]	WHO Guideline [b]
Whole basin							
pH		7.63	8.87	8.35	0.27	6.5–8.5	6.5–8.5
EC	µS/cm	128	738	403	117	—	—
T	°C	8.6	14.7	10.8	1.4	—	—
DO	mg/L	0.40	9.38	8.18	1.52	—	—
Na^+	mmol/L	0.06	0.78	0.30	0.17	—	—
K^+	mmol/L	0.02	0.19	0.05	0.03	—	—
Mg^{2+}	mmol/L	0.08	1.56	0.48	0.26	—	—
Ca^{2+}	mmol/L	0.44	2.79	1.54	0.46	—	—
F^-	mmol/L	0.00	0.02	0.01	0.00	0.05	0.08
Cl^-	mmol/L	0.04	0.55	0.18	0.14	7.05	7.05
NO_3^-	mmol/L	0.002	0.750	0.193	0.130	1.428	3.570
SO_4^{2-}	mmol/L	0.20	2.64	0.85	0.41	2.60	2.60
HCO_3^-	mmol/L	0.52	5.49	2.22	0.79	—	—
NH_4^{+} [a]	mmol/L	0.000	0.010	0.003	—	0.036	0.107
$\delta^{34}S\text{-}SO_4^{2-}$	‰	−7.79	22.13	4.68	6.21	—	—
SAR		0.04	0.61	0.21	0.11	—	—
Na%		1.27	19.02	6.98	3.41	—	—
RSC		−5.26	−0.49	−1.81	0.88	—	—
Main stream							
pH		7.63	8.87	8.36	0.28		
EC	µS/cm	381	480	428	32		
T	°C	9.5	14.7	11.0	1.4		
DO	mg/L	6.85	9.35	8.34	0.67		
Na^+	mmol/L	0.06	0.78	0.30	0.18		
K^+	mmol/L	0.03	0.07	0.05	0.01		
Mg^{2+}	mmol/L	0.34	0.90	0.52	0.13		
Ca^{2+}	mmol/L	1.13	1.96	1.62	0.23		
F^-	mmol/L	0.00	0.01	0.01	0.00		
Cl^-	mmol/L	0.05	0.55	0.19	0.15		
NO_3^-	mmol/L	0.002	0.750	0.207	0.156		
SO_4^{2-}	mmol/L	0.54	1.35	0.89	0.19		
HCO_3^-	mmol/L	1.47	3.09	2.38	0.38		
$\delta^{34}S\text{-}SO_4^{2-}$	‰	−7.79	14	2.84	5.36		
SAR		0.04	0.61	0.21	0.14		
Na%		1.27	19.02	6.53	4.31		
RSC		−2.94	−0.79	−1.92	0.48		

Table 1. *Cont.*

	Unit	Min	Max	Mean	SD	Chinese Guideline [b]	WHO Guideline [b]
Tributaries							
pH		7.75	8.75	8.34	0.27		
EC	μS/cm	128	738	379	160		
T	°C	8.6	13.1	10.5	1.3		
DO	mg/L	0.40	9.38	8.03	2.04		
Na^+	mmol/L	0.12	0.75	0.29	0.16		
K^+	mmol/L	0.02	0.19	0.05	0.04		
Mg^{2+}	mmol/L	0.08	1.56	0.42	0.35		
Ca^{2+}	mmol/L	0.44	2.79	1.44	0.61		
F^-	mmol/L	0.00	0.02	0.01	0.00		
Cl^-	mmol/L	0.04	0.53	0.17	0.13		
NO_3^-	mmol/L	0.017	0.422	0.178	0.094		
SO_4^{2-}	mmol/L	0.20	2.64	0.82	0.57		
HCO_3^-	mmol/L	0.52	5.49	2.04	1.07		
$\delta^{34}S\text{-}SO_4^{2-}$	‰	−6.33	22.13	6.51	6.59		
SAR		0.11	0.39	0.21	0.07		
Na%		4.08	10.77	7.48	2.02		
RSC		−5.26	−0.49	−1.68	1.18		

Note: [a] the data of concentration is from [23]; [b] the unit of related values in Chinese guideline and WHO guideline are converted to mmol/L.

The major chemical ions of the Chishui River water are presented in Figure 2. It is very distinct that the principal cation in Chishui River is Ca^{2+}, which accounts for $71 \pm 6\%$ (mean $\pm$ SD) of the cations, followed by Mg^{2+} ($21 \pm 6\%$) and $Na^+ + K^+$ ($8 \pm 3\%$). The anions sequence is HCO_3^- ($55 \pm 9\%$) > SO_4^{2-} ($41 \pm 9\%$) > Cl^- ($4 \pm 3\%$). HCO_3^- is therefore the predominant anion in Chishui River, which is a widely confirmed product (as well as riverine Ca^{2+} and Mg^{2+}) of CO_2–associated weathering process without the influence of anthropogenic pollution [13,24]. The massive carbonate sedimentary rock distribution in such a karst landscape river watershed creates the advantageous conditions for the chemical weathering of Ca/Mg–contented rock. Although the lower reaches flow through the area where the siliciclastic sedimentary rock developed, the inheritance of river water will make the lower reach reveal the characteristics of dissolved loads from upper and middle streams. Generally, the weathering rate of carbonate rock is much higher than other rocks in the same situations [25,26], implying that the HCO_3^-, Ca^{2+}, and Mg^{2+}, derived from carbonate sedimentary rock weathering and exported into Chishui River in upper–middle stream, are far more than the potential Na^+ and K^+ originated from the weathering of siliciclastic sedimentary rock in the lower stream. Therefore, the hydrochemical type of Chishui River water is a Ca–Mg–HCO_3^- type controlled by carbonate weathering, which is highly in agreement with the studies of the whole Changjiang River (Yangtze River) basin (Figure 2).

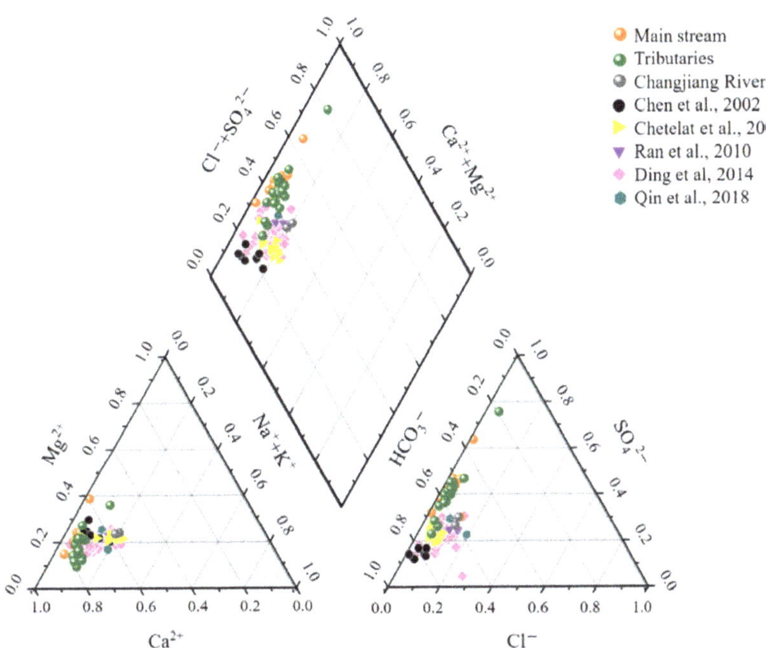

Figure 2. Piper diagrams representing the percentages of hydrochemical species in Chishui River (including 2 water samples collected in Changjiang River before and after the Chishui River drained into, the solid gray circle) and Changjiang River (data reported by previous studies [10,27–30]).

The potential sources of riverine SO$_4$$^{2-}$ mainly include weathering process, atmospheric precipitation, industrial/mining activities, and sewage inputs [10]. Chemical weathering related studies have found that the accelerated carbonate dissolution occurred with relatively high SO$_4$$^{2-}$ concentrations (as an agency of weathering processes) in the karst region [31,32]. Thus, the considerable fraction of SO$_4$$^{2-}$ (41 $\pm$ 9%, close to HCO$_3$$^-$) is non–negligible.

3.2. Source of Fluvial Solutes

3.2.1. Stoichiometry–Revealed Sources of Solutes

In hydrochemistry studies, the high Cl$^-$ concentration could be applied as a significant index of anthropogenic inputs (mainly in urban area) and domestic sewage [33]. Generally, the river nitrate originated from agricultural synthetic fertilizers exhibits a high NO$_3$$^-$ concentration and a high value of NO$_3$$^-$/Cl$^-$ ratio, while the domestic sewage presents low NO$_3$$^-$/Cl$^-$ ratio and high concentration of Cl$^-$ due to highly organic matter [34]. As shown in Figure 3a, the high NO$_3$$^-$/Cl$^-$ ratio and low level of Cl$^-$ concentration (range 0.04~0.55 meq/L, mean value 0.18 $\pm$ 0.14 meq/L) were observed in all water samples of Chishui River, suggesting the major contribution by agricultural input (fertilizer), which can be supported by the large distribution area of cropland (~19.6%) in the whole watershed [7]. By contrast, the potential Cl$^-$ inputs from the anthropogenic domestic sewage in urban region may be assuaged by various landscape setting and the well–mixed processes of river water, further resulting in a low concentration of riverine Cl$^-$. Moreover, although previous work have revealed that rainwater NO$_3$$^-$ could be a significant contributor (even up to 71%) of riverine NO$_3$$^-$ during storm–frequency season [35], however, the river water NO$_3$$^-$ concentration (0.19 $\pm$ 0.13 meq/L) in the study period (dry season) was much higher than the karst rainwater NO$_3$$^-$ concentration (typically < 0.04 meq/L) [36–38]. Therefore, the contribution of wet deposition to Chishui River water NO$_3$$^-$ is very limited.

In addition, the river water SO_4^{2-}/Ca^{2+} ratios were much higher than the NO_3^-/Ca^{2+} ratios (Figure 3b), indicating an additional potential impact of industrial activities and mining (e.g., pyrite in middle reaches) on fluvial solutes [39,40].

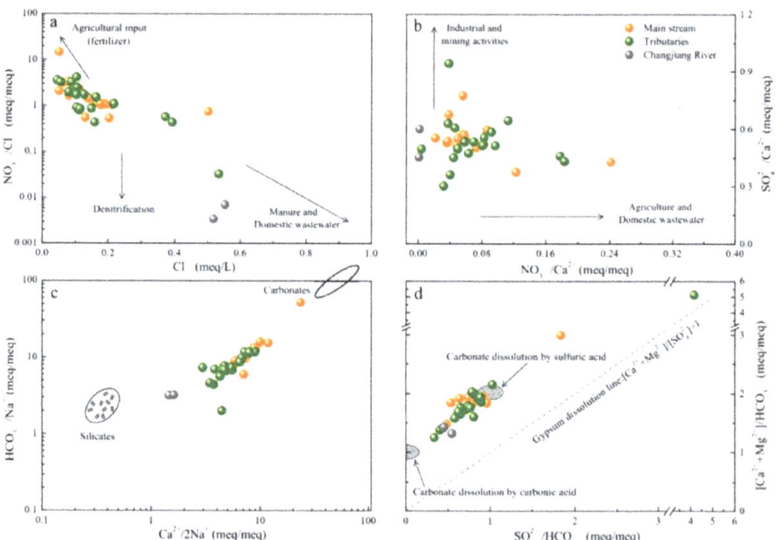

Figure 3. The relationship between Cl^- and NO_3^-/Cl^- ratios (**a**), NO_3^-/Ca^{2+} ratios and SO_4^{2-}/Ca^{2+} ratios (**b**), $Ca^{2+}/2Na^+$ ratios and HCO_3^-/Na^+ ratios (**c**), $[Ca^{2+} + Mg^{2+}]/HCO_3^-$ ratios and SO_4^{2-}/HCO_3^- ratios (**d**) in Chishui River watershed.

The weathering–derived major ions always present typical ratios, thus the ratios of Ca^{2+}/Na^+ and HCO_3^-/Na^+ are the useful index to track the weathering origins of ions [13,19]. The Ca^{2+}/Na^+ and HCO_3^-/Na^+ ratios of Chishui River water were scattered between the silicate and carbonate end–members and more inclined to the carbonate source (Figure 3c), implying the leading role of carbonate weathering products on riverine ions with a slightly mixing process of silicate weathering products. It is noteworthy that the two samples form Changjiang River were more trended to silicate end–members (Figure 3c), indicating the differentiated weathering contributions of various rocks in upper Changjiang River.

According to the chemical reaction equation between carbonate minerals $(Ca_xMg_{(1-x)}CO_3)$ and H_2SO_4 or H_2CO_3 $(H_2O + CO_2)$, the relationships between the $[Ca^{2+} + Mg^{2+}]/HCO_3^-$ ratios and SO_4^{2-}/HCO_3^- ratios could reveal specific weathering process. If the $Ca_xMg_{(1-x)}CO_3$ are only weathered by H_2CO_3, the SO_4^{2-}/HCO_3^- should be ~0 and the $[Ca^{2+} + Mg^{2+}]/HCO_3^-$ ratio should be ~1. However, when only sulfuric acid is involved in chemical weathering of carbonate minerals, the $SO_4^{2-}/HCO_3^- \approx 1$ and $[Ca^{2+} + Mg^{2+}]/HCO_3^- \approx 2$. In Figure 3d, most of samples distributed between the two types of acid–involved weathering results, implying that both H_2SO_4 and H_2CO_3 weathering processes occurred. The samples were relatively closer to the H_2SO_4–involved weathering results, indicating the importance of sulfuric acid weathering processes, which may notably accelerate the rock dissolution. The generally low Mg^{2+}/Ca^{2+} ratios (<0.5) of most river water samples further distinguished the contributions of different types of carbonate minerals, that is, the riverine Ca^{2+} and Mg^{2+} were primarily controlled by calcite dissolution (the dolomite dissolution dominated river water $Mg^{2+}/Ca^{2+} \approx 1$) [26]. In contrast, if the extensive silicate weathering occurs, the $[Ca^{2+} + Mg^{2+}]/HCO_3^-$ ratios will decreased and the sample points in Figure 3d will move downwards systematically due to the simultaneously production of Na^+, K^+, HCO_3^-, and SO_4^{2-} in silicate weathering [11,41].

However, there were no samples with $[Ca^{2+} + Mg^{2+}]/HCO_3^- < 1$ found in Chishui River, further reflecting the negligibility of silicate weathering contribution. In addition, all samples were completely above the gypsum dissolution line ($[Ca^{2+} + Mg^{2+}]/SO_4^{2-} = 1$) (Figure 3d), demonstrating the limited contribution of gypsum dissolution to riverine solutes [19,42]. Similar to H_2SO_4, the nitric acid (HNO_3) could also accelerate the weathering processes via dissolving the minerals [1], particularly in a small karst watershed scale with high nitrate loads [15,17,18]. Given the NO_3^- concentration much lower than HCO_3^- and SO_4^{2-} concentration (Table 1), here we concluded that the HNO_3–related rock weathering was also relatively limited.

According to low concentrations of riverine K^+ (0.05 mmol/L), F^- (0.01 mmol/L), and NH_4^+ (0.003 mmol/L) [23], these three ions related ratios were not presented in Figure 3. Even so, previous studies have concluded that K^+ is contributed both by rock weathering (silicate) and human emissions [10,41,43]. Thus, here we hold the opinion that K^+ is mainly controlled by rock source in such a low riverine K^+ level (anthropogenic input is negligible) in Chishui River. Riverine F^- is also primarily originated from bedrocks [44]. As for NH_4^+, the agriculture–related processes, mainly ammonia–contained fertilizer, can be an interpretation for the samples with relatively high ammonia concentration [45].

3.2.2. PCA Analysis

The PCA was used to further explore the potential sources of riverine ions in Chishui River. Three principal components (PCs) with eigenvalues exceeding 1 were distinguished and illustrated in Figure 4, with more details presented in Table 2. These three PCs contained a total of 79.67% variance. The PC 1, PC 2, and PC 3 accounted for 33.8%, 23.9%, and 23.1% of total variances, respectively. Four ions (K^+, Mg^{2+}, F^-, and HCO_3^-) as well as EC presented notably positive loadings in PC 1, and the relatively strong loading of Ca^{2+} (0.48) is also found in PC 1 (Figure 4 and Table 2). Based on the stoichiometry discussed in Section 3.2.1, here we infer that PC 1 was mainly contributed by natural rock weathering inputs in the Chishui River watershed due to the high loadings of feature ions of rock weathering (primarily carbonate) [46]. Na^+ and Cl^- presented clear positive loadings in PC 2, but these two ions were weakly loaded in PC 1 (natural source) and PC 3 (SO_4^{2-}—dominated human inputs). Combined with the negligible contribution of silicate weathering and the limited anthropogenic Cl^- inputs mentioned before, the atmospheric input/deposition could be a reasonable contributor, which is also supported by previous studies [42]. The strongest loadings of SO_4^{2-} and NO_3^- were observed in PC 3 (Figure 4), considering as typical anthropogenic inputs. It is noteworthy that the PC 3 also presented a significantly positive loading of Ca^{2+}, further supporting the H_2SO_4–involved weathering processes. Additionally, the same positive loading of EC (reflect total riverine ion content) was found in PC 1 and PC 3 (Table 2), indicating that the contribution of natural and anthropogenic sources to total fluvial ions was comparable.

3.3. Sulfur Isotope–Based Source Identification of Riverine Sulfate

Although limited impact of fluvial SO_4^{2-} for drinking and irrigation purpose is reported, the potential accelerated weathering may result in more solutes input to river system, which further causes other environmental issues (e.g., carbon source/sink variations) [11]. Therefore, to further constrain the SO_4^{2-} sources, the effective indicator $\delta^{34}S$–SO_4^{2-} was applied. Generally, the riverine SO_4^{2-} is derived from four sources: atmospheric inputs (mainly rainwater); sulfide oxidation (e.g., pyrite); human inputs (e.g., agricultural fertilizers, sewage, and industrial effluent); gypsum dissolution (a kind of evaporitic rock) [12,14,47]. These four sources can be well constrained by combining the distinguished $\delta^{34}S$–SO_4^{2-} values and SO_4^{2-} concentrations (Figure 5).

The sulfur isotope compositions of atmospheric depositions in south China were well explored in previous studies, such as the 950–day based study in Wuhan urban area ($\delta^{34}S$–SO_4^{2-} = +4.5 $\pm$ 1.3‰) [48], rainwater in Sichuan Basin ($\delta^{34}S$–SO_4^{2-} = +1.5~ + 6.0‰) [49], and the rainwater sulfur isotope study carried out in a karst agricultural catchment (δ^{34}S–SO$_4{}^{2-}$ = +1.3 ± 6.2‰) [50]. Based on the similar geographical location, climatic condition, lithology and agriculture–dominated industrial structure, the rainwater δ^{34}S–SO$_4{}^{2-}$ values (+1.3 ± 6.2‰) of the karst agricultural catchment can be regarded as an atmospheric input end–member. The sulfide oxidation derived SO$_4{}^{2-}$ usually inherit the sulfur isotope compositions of the initial sulfide due to limited isotope fractionation during oxidation processes [19]. The average δ^{34}S–SO$_4{}^{2-}$ value of pyrites in the study area (Sichuan, Guizhou, and Yunnan provinces) was −7.8‰ [51], which is within the typical range of sulfide (−13.0~−1.1‰) [19]. Moreover, although the sulfide oxidated SO$_4{}^{2-}$ concentrations were rarely reported in the study area, previous work have shown that its concentration should be exceed 0.1 mol/L [47], even up to 2 mol/L, that is, 1/[SO$_4{}^{2-}$] = 0.5~10. The anthropogenic inputs, mainly including agricultural fertilizers, domestic and industrial sewage, present a wide range of sulfur isotope compositions. Therefore, the typical δ^{34}S–SO$_4{}^{2-}$ values of water in farmland and livestock farm areas (+2.8~+13.9‰) were regarded as the end–member of human inputs [12,52], and the potentially high SO$_4{}^{2-}$ concentration (1/[SO$_4{}^{2-}$] below ~17.5) [19] was selected to distinguish the overlapped part of the δ^{34}S–SO$_4{}^{2-}$ values with atmospheric input. Moreover, the gypsum dissolution source presented the highest δ^{34}S–SO$_4{}^{2-}$ value (up to ~+30‰) [2] and high SO$_4{}^{2-}$ concentration.

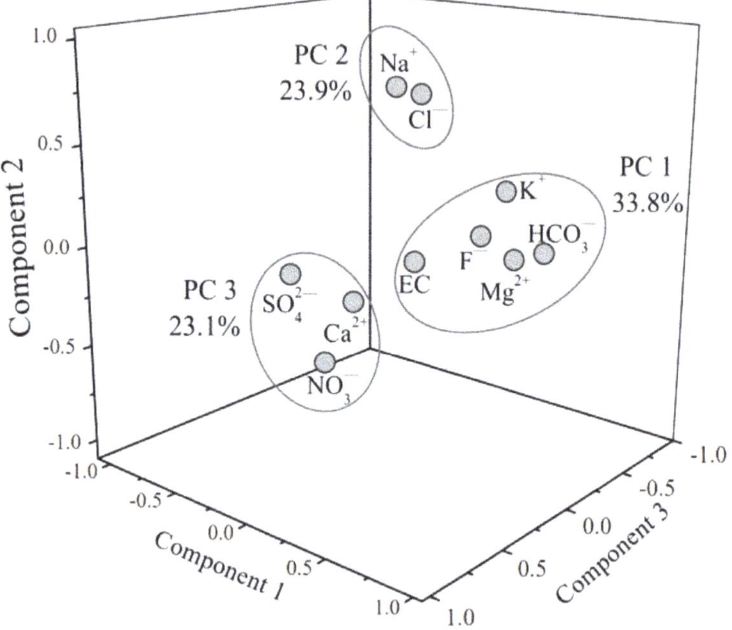

Figure 4. The 3D plot of factor loadings of Chishui River ions and other parameters.

As shown in Figure 5, the ternary diagram presented the relative contributions of different origins to fluvial SO$_4{}^{2-}$ via the relationship between δ^{34}S–SO$_4{}^{2-}$ values and 1/[SO$_4{}^{2-}$], this is, riverine SO$_4{}^{2-}$ were controlled by the sources of human inputs and sulfide oxidation. The middle–stream–distributed pyrite deposits and the relatively high population density (282 persons per km^2) and wide urbanization area (1.2%) [7,23,40] also underpinned these results. It is noteworthy that four samples were distributed over the range of human inputs and tended into the gypsum dissolution sources (Figure 5). This suggests the potential influence of gypsum dissolution on river water in some sampling sites to a certain extent, but it was negligible on a whole watershed scale, which was also supported by previous work [19]. Although the range of $\delta^{34}S$–SO_4^{2-} value of atmospheric inputs was partially overlapped with other sources, the much lower SO_4^{2-} concentrations of atmospheric source were significantly separated from human input and sulfide oxidation sources (Figure 5), confirming again that the contribution of atmospheric inputs to riverine sulfate was very limited. This can also be supported by the average investigation results of global rivers (atmospheric input contributed to ~3% sulfate) [13]. In comparison, the water samples obtained in larger–scale Zhujiang River watershed (originated in karst region of Guizhou province) in Figure 5 presented a distribution closer to the sulfide oxidation end–member, with some samples trended to the atmospheric source [11], indicating a complicated mixing processes and differential contributions of sulfate sources in different scaled catchments.

Table 2. Varimax rotated component matrix for water ions in Chishui River.

Eigenvalues	5.78	2.03	1.07	Communalities
Variance (%)	33.80	23.90	23.10	
Cumulative (%)	33.80	57.70	80.80	
Variable	PC 1	PC 2	PC 3	
EC	0.69	0.19	0.69	0.97
Na^+	0.27	0.84	0.33	0.88
K^+	0.75	0.40	0.14	0.75
Mg^{2+}	0.91	0.14	0.28	0.92
Ca^{2+}	0.48	0.01	0.85	0.95
F^-	0.70	0.21	0.26	0.61
Cl^-	0.33	0.80	0.24	0.81
NO_3^-	−0.13	−0.54	0.40	0.47
SO_4^{2-}	0.19	0.09	0.92	0.90
HCO_3^-	0.94	0.14	0.10	0.92

Note: Extraction method: Principal component analysis; Rotation method: Varimax with Kaiser normalization; The significance of KMO and Bartlett's sphericity test is <0.001.

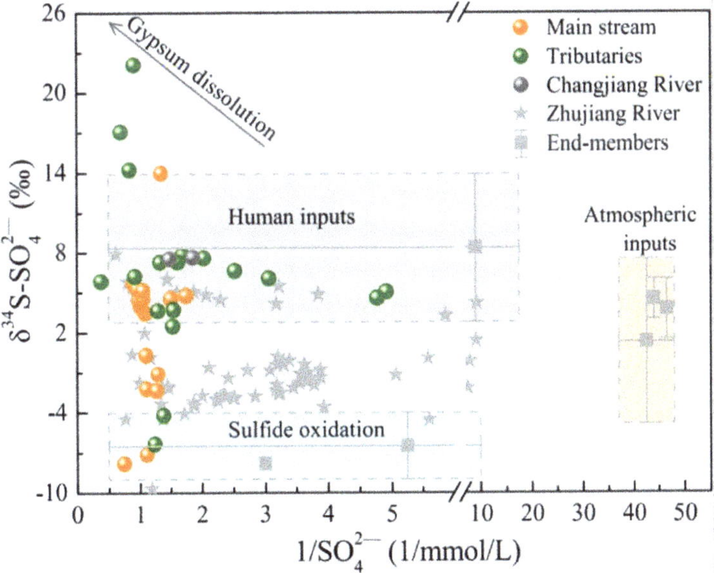

Figure 5. The relationship between $\delta^{34}S$–SO_4^{2-} values and SO_4^{2-} concentrations revealing the mixing processes of riverine sulfate from different sources. The data of Zhujiang River is from [11].

3.4. Water Quality and Risk Assessment

3.4.1. Irrigation and Guideline–Based Water Quality

Chishui River servers as primary water resources of the agriculture, industry, and local residents (population > 5 million). As summarized in Table 1, according to the Chinese and WHO drinking water quality guidelines, the desirable pH value for drinking is recommended between 6.5~8.5, thus most of the water samples (81.6%) had suitable pH values (Table S1). The high pH values (>8.5) of partial samples were mainly caused by carbonate weathering. Most of the guidelines–covered ions, including F^-, Cl^-, NO_3^-, and NH_4^+ of all river water samples were below the recommended limits. However, although the average value of riverine SO_4^{2-} (0.85 mmol/L) was lower than that of guidelines limits (2.60 mmol/L), the maximum value of riverine SO_4^{2-} (2.64 mmol/L) exceeded the permissible limits, which needs more attention. The previous study on trace metal also assessed the 19 metals–based water quality index (WQI) of Chishui River, which showed the assessment results of excellent water quality (WQI < 50, only from the perspective of metals) without extensive trace metal contamination [7].

The Na% and SAR indicators can reflect the Na hazard to agricultural land by irrigation–influenced soil aggregate [19]. According to the calculated SAR and Na% values, all water samples in the Chishui River could be regarded as excellent/good quality (SAR < 0.61; Na% < 19.0%) with no samples out of the desirable limits (SAR = 1; Na% = 30.0%). For the residual sodium carbonate (RSC), there was no water samples present a RSC value exceeded 1.25 (Table S1), indicating the well suitability of irrigation water. Moreover, the United States Salinity Labortory (USSL) diagram and Wilcox diagram were plotted based on the EC, SAR, and Na% values (Figure 6), and all the Chishui River water samples were scattered in the C1S1 and C2S1 zone of USSL diagram and the "Excellent to Good' area of Wilcox diagram. Overall, the Chishui River water is desirable for agricultural irrigation and will not bring about the soil hazard. However, the continuous long–term monitoring is still a worthy mission due to the enhanced human activities.

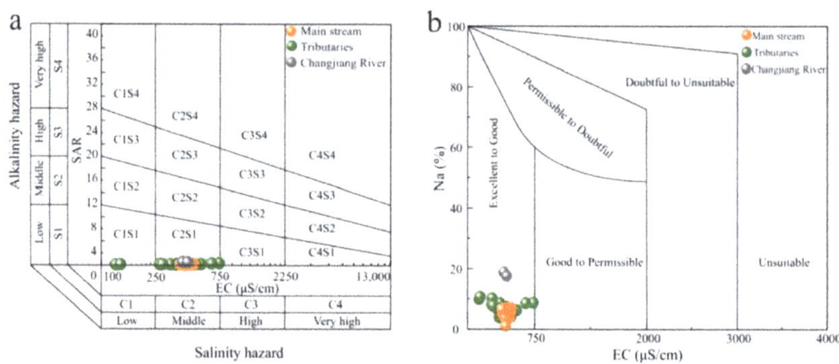

Figure 6. Irrigation water quality assessment alkalinity and salinity: (**a**) United States Salinity Labortory (USSL) diagram and (**b**) Wilcox diagram.

3.4.2. Health Risk Assessment

Among the major ions in natural water, excessive exposure (ingestion intake) of F^-, NO_3^-, and NH_4^+ can cause typical non–carcinogenic hazards, while SO_4^{2-} does not result in health issue [11]. Thus, riverine F^-, NO_3^-, and NH_4^+ were involved in the health risk assessment in this study. The average concentration–based HQ values were calculated and presented in Figure 7. The HQ values of three assessed ions were in the order of HQ–NO_3^- (0.28 ± 0.19) > HQ–F^- (0.13 ± 0.06) > HQ–NH_4^+ (0.002 ± 0.000) for children and HQ–NO_3^- (0.11 ± 0.07) > HQ–F^- (0.05 ± 0.02) > HQ–NH_4^+ (0.001 ± 0.000) for adult. The results were similar to other researches [44], that is, NH_4^+ occupied a tiny part of the total HQ, while NO_3^- and F^- were the majority. For both children and adults, HQs of each ion and total HQ were all lower than the hazard level (<1), indicating a lower health risk of assessed ions. It is noteworthy that the HQ value for children was always higher than that for adults, implying that children were facing higher risks from riverine ions compared to adults. In addition, the potential harmful health effects can also occur if the HQ value is greater than 0.1 for children [21]. Therefore, the riverine ion–related health risk is not completely negligible, particularly for children.

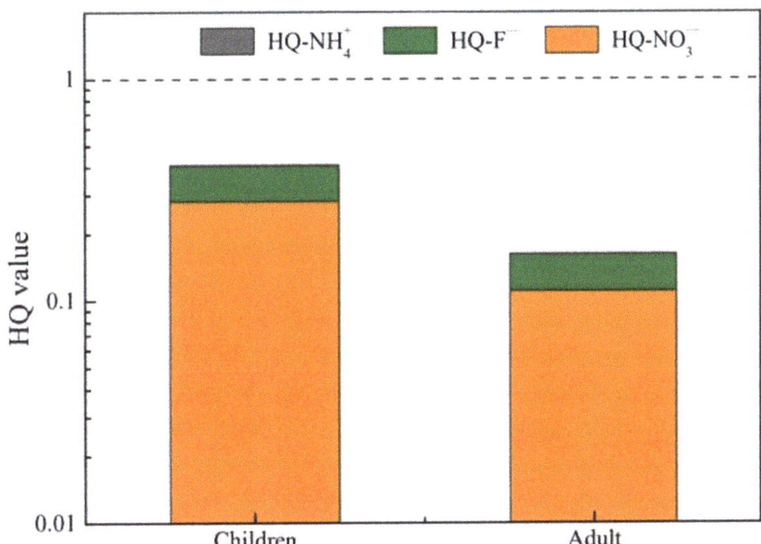

Figure 7. Non–carcinogenic health risk assessment of fluorine, nitrate, ammonia in Chishui River water.

4. Conclusions

The present study aimed to investigate hydrochemical compositions and sulfur isotope compositions of Chishui River water, southwest of China. Based on the isotope tracer, stoichiometry and PCA method, the sources of major ions were identified. The rock (mainly carbonate) weathering inputs were the primary sources of K^+, Mg^{2+}, Ca^{2+}, F^-, HCO_3^-, and atmospheric contribution was the source of Na^+ and Cl^-, while the anthropogenic input was responsible for SO_4^{2-} and NO_3^-. Sulfur isotope compositions further reflected the relative contribution of different sources, that is, the contribution of human inputs (main) and sulfide oxidation controlled the riverine SO_4^{2-}, but the input of atmospheric deposition was limited. Both the water quality and hazard quotient assessment produced good results, indicating the suitable aims of river water for drinking and irrigation. This work and the possible continuing study would be benefit to the prevention of watershed water environment and further help planning the sustainable water resources in Chishui River watershed.

Supplementary Materials: The following are available online at https://www.mdpi.com/article/10.3390/w13091231/s1, Table S1: The water parameters, major ion compositions, $\delta^{34}S$–SO_4^{2-} values, SAR, Na%, RSC, $[TZ^+ - TZ^-]/TZ^-$ of Chishui River.

Author Contributions: Conceptualization, Q.W. and X.G.; methodology, Q.W. and X.G.; software, X.G.; validation, Q.W.; formal analysis, X.G., Q.W., Z.W., S.G. and T.W.; investigation, Q.W.; resources, Q.W. and Z.W.; writing—original draft preparation, X.G. and Q.W.; writing—review and editing, X.G., Q.W. and Z.W.; supervision, Q.W.; project administration, Q.W. and Z.W.; funding acquisition, Q.W. and Z.W. All authors have read and agreed to the published version of the manuscript.

Funding: This research was funded by the Joint Fund of the National Natural Science Foundation of China and Guizhou Province, China (U1612442), the National Natural Science Foundation of China (41863004, 41863003, 41763019), the first–class discipline construction project in Guizhou Province–Public Health and Preventive Medicine (No. 2017[85], GNYL [2017]007), and the Guizhou Science and Technology Support Program ([2019]2832). The funders had no role in study design, data collection and analysis, decision to publish, or preparation of the manuscript.

Institutional Review Board Statement: Not applicable.

Informed Consent Statement: Not applicable.

Data Availability Statement: The data are presented in this study are available in Supplementary Materials.

Conflicts of Interest: The authors declare no conflict of interest.

References

1. Barnes, R.T.; Raymond, P.A. The contribution of agricultural and urban activities to inorganic carbon fluxes within temperate watersheds. *Chem. Geol.* **2009**, *266*, 318–327. [CrossRef]
2. Nightingale, M.; Mayer, B. Identifying sources and processes controlling the sulphur cycle in the Canyon Creek watershed, Alberta, Canada. *Isot. Environ. Health Stud.* **2012**, *48*, 89–104. [CrossRef]
3. Zeng, J.; Han, G. Preliminary copper isotope study on particulate matter in Zhujiang River, southwest China: Application for source identification. *Ecotoxicol. Environ. Saf.* **2020**, *198*, 110663. [CrossRef] [PubMed]
4. Wang, J.; Jiang, Y.; Sun, J.; She, J.; Yin, M.; Fang, F.; Xiao, T.; Song, G.; Liu, J. Geochemical transfer of cadmium in river sediments near a lead-zinc smelter. *Ecotoxicol. Environ. Saf.* **2020**, *196*, 110529. [CrossRef] [PubMed]
5. Zaric, N.M.; Deljanin, I.; Ilijević, K.; Stanisavljević, L.; Ristić, M.; Gržetić, I. Assessment of spatial and temporal variations in trace element concentrations using honeybees (*Apis mellifera*) as bioindicators. *PeerJ* **2018**, *6*, e5197. [CrossRef] [PubMed]
6. Wang, J.; Wang, L.; Wang, Y.; Tsang, D.C.W.; Yang, X.; Beiyuan, J.; Yin, M.; Xiao, T.; Jiang, Y.; Lin, W.; et al. Emerging risks of toxic metal(loid)s in soil-vegetables influenced by steel-making activities and isotopic source apportionment. *Environ. Int.* **2021**, *146*, 106207. [CrossRef]
7. Xu, S.; Lang, Y.; Zhong, J.; Xiao, M.; Ding, H. Coupled controls of climate, lithology and land use on dissolved trace elements in a karst river system. *J. Hydrol.* **2020**, *591*, 125328. [CrossRef]
8. Liu, M.; Han, G.; Zhang, Q. Effects of agricultural abandonment on soil aggregation, soil organic carbon storage and stabilization: Results from observation in a small karst catchment, Southwest China. *Agric. Ecosyst. Environ.* **2020**, *288*, 106719. [CrossRef]
9. Chen, L.; Liu, J.-r.; Hu, W.-f.; Gao, J.; Yang, J.-y. Vanadium in soil-plant system: Source, fate, toxicity, and bioremediation. *J. Hazard. Mater.* **2021**, *405*, 124200. [CrossRef]
10. Chetelat, B.; Liu, C.Q.; Zhao, Z.Q.; Wang, Q.L.; Li, S.L.; Li, J.; Wang, B.L. Geochemistry of the dissolved load of the Changjiang Basin rivers: Anthropogenic impacts and chemical weathering. *Geochim. Cosmochim. Acta* **2008**, *72*, 4254–4277. [CrossRef]
11. Liu, J.; Han, G. Distributions and Source Identification of the Major Ions in Zhujiang River, Southwest China: Examining the Relationships Between Human Perturbations, Chemical Weathering, Water Quality and Health Risk. *Exposure Health* **2020**, *12*, 849–862. [CrossRef]
12. Szynkiewicz, A.; Witcher, J.C.; Modelska, M.; Borrok, D.M.; Pratt, L.M. Anthropogenic sulfate loads in the Rio Grande, New Mexico (USA). *Chem. Geol.* **2011**, *283*, 194–209. [CrossRef]
13. Gaillardet, J.; Dupré, B.; Louvat, P.; Allègre, C.J. Global silicate weathering and CO2 consumption rates deduced from the chemistry of large rivers. *Chem. Geol.* **1999**, *159*, 3–30. [CrossRef]
14. Xu, S.; Li, S.; Su, J.; Yue, F.; Zhong, J.; Chen, S. Oxidation of pyrite and reducing nitrogen fertilizer enhanced the carbon cycle by driving terrestrial chemical weathering. *Sci. Total Environ.* **2021**, *768*, 144343. [CrossRef]
15. Zeng, J.; Yue, F.-J.; Wang, Z.-J.; Wu, Q.; Qin, C.-Q.; Li, S.-L. Quantifying depression trapping effect on rainwater chemical composition during the rainy season in karst agricultural area, southwestern China. *Atmos. Environ.* **2019**, *218*, 116998. [CrossRef]
16. Han, G.; Tang, Y.; Liu, M.; Van Zwieten, L.; Yang, X.; Yu, C.; Wang, H.; Song, Z. Carbon-nitrogen isotope coupling of soil organic matter in a karst region under land use change, Southwest China. *Agric. Ecosyst. Environ.* **2020**, *301*, 107027. [CrossRef]
17. Wang, Z.-J.; Li, S.-L.; Yue, F.-J.; Qin, C.-Q.; Buckerfield, S.; Zeng, J. Rainfall driven nitrate transport in agricultural karst surface river system: Insight from high resolution hydrochemistry and nitrate isotopes. *Agric. Ecosyst. Environ.* **2020**, *291*, 106787. [CrossRef]
18. Yue, F.-J.; Waldron, S.; Li, S.-L.; Wang, Z.-J.; Zeng, J.; Xu, S.; Zhang, Z.-C.; Oliver, D.M. Land use interacts with changes in catchment hydrology to generate chronic nitrate pollution in karst waters and strong seasonality in excess nitrate export. *Sci. Total Environ.* **2019**, *696*, 134062. [CrossRef]
19. Liu, J.; Han, G. Major ions and δ34SSO4 in Jiulongjiang River water: Investigating the relationships between natural chemical weathering and human perturbations. *Sci. Total Environ.* **2020**, *724*, 138208. [CrossRef]
20. Cao, X.; Wu, P.; Zhou, S.; Sun, J.; Han, Z. Tracing the origin and geochemical processes of dissolved sulphate in a karst-dominated wetland catchment using stable isotope indicators. *J. Hydrol.* **2018**, *562*, 210–222. [CrossRef]

21. Zeng, J.; Han, G.; Yang, K. Assessment and sources of heavy metals in suspended particulate matter in a tropical catchment, northeast Thailand. *J. Clean. Prod.* **2020**, *265*, 121898. [CrossRef]
22. Gaillardet, J.; Dupre, B.; Allegre, C.J.; Négrel, P. Chemical and physical denudation in the Amazon River Basin. *Chem. Geol.* **1997**, *142*, 141–173. [CrossRef]
23. An, Y.; Jiang, H.; Wu, Q.; Yang, R.; Lang, X.; Luo, J. Assessment of water quality in the Chishui River Basin during low water period. *Resour. Environ. Yangtze Basin* **2014**, *23*, 1472–1478. (In Chinese)
24. Liu, J.; Han, G. Controlling factors of riverine CO_2 partial pressure and CO_2 outgassing in a large karst river under base flow condition. *J. Hydrol.* **2021**, *593*, 125638. [CrossRef]
25. Tipper, E.T.; Bickle, M.J.; Galy, A.; West, A.J.; Pomiès, C.; Chapman, H.J. The short term climatic sensitivity of carbonate and silicate weathering fluxes: Insight from seasonal variations in river chemistry. *Geochim. Cosmochim. Acta* **2006**, *70*, 2737–2754. [CrossRef]
26. Zeng, J.; Yue, F.-J.; Li, S.-L.; Wang, Z.-J.; Wu, Q.; Qin, C.-Q.; Yan, Z.-L. Determining rainwater chemistry to reveal alkaline rain trend in Southwest China: Evidence from a frequent-rainy karst area with extensive agricultural production. *Environ. Pollut.* **2020**, *266*, 115166. [CrossRef] [PubMed]
27. Chen, J.; Wang, F.; Xia, X.; Zhang, L. Major element chemistry of the Changjiang (Yangtze River). *Chem. Geol.* **2002**, *187*, 231–255. [CrossRef]
28. Ran, X.; Yu, Z.; Yao, Q.; Chen, H.; Mi, T. Major ion geochemistry and nutrient behaviour in the mixing zone of the Changjiang (Yangtze) River and its tributaries in the Three Gorges Reservoir. *Hydrol. Process.* **2010**, *24*, 2481–2495. [CrossRef]
29. Tiping, D.; Gao, J.; Tian, S.; Shi, G.; Chen, F.; Wang, C.; Luo, X.; Han, D. Chemical and Isotopic Characteristics of the Water and Suspended Particulate Materials in the Yangtze River and Their Geological and Environmental Implications. *Acta Geol. Sin. Engl. Ed.* **2014**, *88*, 276–360. [CrossRef]
30. Qin, T.; Yang, P.; Groves, C.; Chen, F.; Xie, G.; Zhan, Z. Natural and anthropogenic factors affecting geochemistry of the Jialing and Yangtze Rivers in urban Chongqing, SW China. *Appl. Geochem.* **2018**, *98*, 448–458. [CrossRef]
31. Li, S.-L.; Calmels, D.; Han, G.; Gaillardet, J.; Liu, C.-Q. Sulfuric acid as an agent of carbonate weathering constrained by δ13CDIC: Examples from Southwest China. *Earth Planet. Sci. Lett.* **2008**, *270*, 189–199. [CrossRef]
32. Liu, J.; Han, G. Tracing Riverine Particulate Black Carbon Sources in Xijiang River Basin: Insight from Stable Isotopic Composition and Bayesian Mixing Model. *Water Res.* **2021**, *194*, 116932. [CrossRef]
33. Widory, D.; Petelet-Giraud, E.; Négrel, P.; Ladouche, B. Tracking the Sources of Nitrate in Groundwater Using Coupled Nitrogen and Boron Isotopes: A Synthesis. *Environ. Sci. Technol.* **2005**, *39*, 539–548. [CrossRef] [PubMed]
34. Yue, F.-J.; Li, S.-L.; Waldron, S.; Wang, Z.-J.; Oliver, D.M.; Chen, X.; Liu, C.-Q. Rainfall and conduit drainage combine to accelerate nitrate loss from a karst agroecosystem: Insights from stable isotope tracing and high-frequency nitrate sensing. *Water Res.* **2020**, *186*, 116388. [CrossRef]
35. Divers, M.T.; Elliott, E.M.; Bain, D.J. Constraining Nitrogen Inputs to Urban Streams from Leaking Sewers Using Inverse Modeling: Implications for Dissolved Inorganic Nitrogen (DIN) Retention in Urban Environments. *Environ. Sci. Technol.* **2013**, *47*, 1816–1823. [CrossRef] [PubMed]
36. Zeng, J.; Yue, F.-J.; Li, S.-L.; Wang, Z.-J.; Qin, C.-Q.; Wu, Q.-X.; Xu, S. Agriculture driven nitrogen wet deposition in a karst catchment in southwest China. *Agric. Ecosyst. Environ.* **2020**, *294*, 106883. [CrossRef]
37. Han, G.; Song, Z.; Tang, Y.; Wu, Q.; Wang, Z. Ca and Sr isotope compositions of rainwater from Guiyang city, Southwest China: Implication for the sources of atmospheric aerosols and their seasonal variations. *Atmos. Environ.* **2019**, *214*, 116854. [CrossRef]
38. Qu, R.; Han, G. A critical review of the variation in rainwater acidity in 24 Chinese cities during 1982–2018. *Elem. Sci. Anthr.* **2021**, *9*, 00142. [CrossRef]
39. Zuo, Y.; An, Y.; Wu, Q.; Qu, K.; Fan, G.; Ye, Z.; Qin, L.; Qian, J.; Tu, C. Study on the hydrochemical characteristics of Duliu River basin in Guizhou Province. *China Environ. Sci.* **2017**, *37*, 2684–2690. (In Chinese)
40. Luo, J.; An, Y.; Wu, Q.; Yang, R.; Jiang, H.; Peng, W.; Yu, X.; Lv, J. Spatial distribution of surface water chemical components in the middle and lower reaches of Chishui River Basin. *Earth Environ.* **2014**, *42*, 297–305. (In Chinese)
41. Xu, Z.; Liu, C.-Q. Chemical weathering in the upper reaches of Xijiang River draining the Yunnan–Guizhou Plateau, Southwest China. *Chem. Geol.* **2007**, *239*, 83–95. [CrossRef]
42. Han, G.; Liu, C.-Q. Water geochemistry controlled by carbonate dissolution: A study of the river waters draining karst-dominated terrain, Guizhou Province, China. *Chem. Geol.* **2004**, *204*, 1–21. [CrossRef]
43. Li, X.; Han, G.; Zhang, Q.; Miao, Z. An optimal separation method for high-precision K isotope analysis by using MC-ICP-MS with a dummy bucket. *J. Anal. At. Spectrom.* **2020**, *35*, 1330–1339. [CrossRef]
44. Li, P.; He, X.; Li, Y.; Xiang, G. Occurrence and Health Implication of Fluoride in Groundwater of Loess Aquifer in the Chinese Loess Plateau: A Case Study of Tongchuan, Northwest China. *Exposure Health* **2019**, *11*, 95–107. [CrossRef]
45. Xu, S.; Li, S.-L.; Zhong, J.; Li, C. Spatial scale effects of the variable relationships between landscape pattern and water quality: Example from an agricultural karst river basin, Southwestern China. *Agric. Ecosyst. Environ.* **2020**, *300*, 106999. [CrossRef]
46. Lee, K.-S.; Ryu, J.-S.; Ahn, K.-H.; Chang, H.-W.; Lee, D. Factors controlling carbon isotope ratios of dissolved inorganic carbon in two major tributaries of the Han River, Korea. *Hydrol. Process.* **2007**, *21*, 500–509. [CrossRef]
47. Han, G.; Tang, Y.; Wu, Q.; Liu, M.; Wang, Z. Assessing Contamination Sources by Using Sulfur and Oxygen Isotopes of Sulfate Ions in Xijiang River Basin, Southwest China. *J. Environ. Qual.* **2019**, *48*, 1507–1516. [CrossRef] [PubMed]

48. Li, X.; Bao, H.; Gan, Y.; Zhou, A.; Liu, Y. Multiple oxygen and sulfur isotope compositions of secondary atmospheric sulfate in a mega-city in central China. *Atmos. Environ.* **2013**, *81*, 591–599. [CrossRef]
49. Li, X.-D.; Masuda, H.; Kusakabe, M.; Yanagisawa, F.; Zeng, H.-A. Degradation of groundwater quality due to anthropogenic sulfur and nitrogen contamination in the Sichuan Basin, China. *Geochem. J.* **2006**, *40*, 309–332. [CrossRef]
50. Yan, Z.; Han, X.; Yue, F.-J.; Zhong, J.; Wang, Z.-J.; Zeng, J.; Li, S.-L. Aquatic chemistry and sulfur isotope composition of precipitation in a karstic agricultuaral area, Southwest China. *Earth Environ.* **2019**, *47*, 811–819. (In Chinese)
51. Gan, C. Geological characteristics and genetic investigation of the Maochang type pyrite deposits. *Miner. Depos.* **1985**, *4*, 51–57. (In Chinese)
52. Das, A.; Pawar, N.J.; Veizer, J. Sources of sulfur in Deccan Trap rivers: A reconnaissance isotope study. *Appl. Geochem.* **2011**, *26*, 301–307. [CrossRef]

Article

Ge/Si Ratio of River Water in the Yarlung Tsangpo: Implications for Hydrothermal Input and Chemical Weathering

Yuchen Wang [1,2,3,†], Tong Zhao [1,2,†], Zhifang Xu [1,2,3,*], Huiguo Sun [1,2,3] and Jiangyi Zhang [1,2,3]

1   Key Laboratory of Cenozoic Geology and Environment, Institute of Geology and Geophysics, Chinese Academy of Sciences, Beijing 100029, China; wangyuchen@mail.iggcas.ac.cn (Y.W.); zhaotong@mail.iggcas.ac.cn (T.Z.); shg@mail.iggcas.ac.cn (H.S.); zhangjiangyi@mail.iggcas.ac.cn (J.Z.)
2   CAS Center for Excellence in Life and Paleoenvironment, Beijing 100044, China
3   University of Chinese Academy of Sciences, Beijing 100049, China
*   Correspondence: zfxu@mail.iggcas.ac.cn
†   These authors contributed equally to this work.

Abstract: Germanium/Silicon (Ge/Si) ratio is a common proxy for primary mineral dissolution and secondary clay formation yet could be affected by hydrothermal and anthropogenic activities. To decipher the main controls of riverine Ge/Si ratios and evaluate the validity of the Ge/Si ratio as a weathering proxy in the Tibetan Plateau, a detailed study was presented on Ge/Si ratios in the Yarlung Tsangpo River, southern Tibetan Plateau. River water and hydrothermal water were collected across different climatic and tectonic zones, with altitudes ranging from 800 m to 5000 m. The correlations between TDS (total dissolved solids) and the Ge/Si ratio and Si and Ge concentrations of river water, combined with the spatial and temporal variations of the Ge/Si ratio, indicate that the contribution of hydrothermal water significantly affects the Ge/Si ratio of the Yarlung Tsangpo River water, especially in the upper and middle reaches. Based on the mass balance calculation, a significant amount of Ge (11–88%) has been lost during its transportation from hydrothermal water to the river system; these could result from the incorporation of Ge on/into clays, iron hydroxide, and sulfate mineral. In comparison, due to the hydrothermal input, the average Ge/Si ratio in the Yarlung Tsangpo River is a magnitude order higher than the majority of rivers over the world. Therefore, evaluation of the contribution of hydrothermal sources should be considered when using the Ge/Si ratio to trace silicate weathering in rivers around the Tibetan Plateau.

Keywords: Ge/Si ratio; Yarlung Tsangpo; Tibetan Plateau; hydrothermal input; silicate weathering

Citation: Wang, Y.; Zhao, T.; Xu, Z.; Sun, H.; Zhang, J. Ge/Si Ratio of River Water in the Yarlung Tsangpo: Implications for Hydrothermal Input and Chemical Weathering. *Water* **2022**, *14*, 181. https://doi.org/10.3390/w14020181

Academic Editor: David Widory

Received: 5 December 2021
Accepted: 6 January 2022
Published: 10 January 2022

Publisher's Note: MDPI stays neutral with regard to jurisdictional claims in published maps and institutional affiliations.

1. Introduction

Chemical weathering regulates atmospheric CO_2 and the Earth's habitability on a long-term scale [1,2]. Since Raymo and Ruddiman proposed the role of tectonic uplift on silicate weathering and global climate cooling during the Cenozoic era [3], the Tibetan Plateau has received considerable attention [4–9]. Given that rivers integrate dissolved and solid weathering products, river geochemistry studies could shed light on weathering processes and fluxes at the basin/global scale [5,10–15].

Ge/Si ratios have been widely used as a silicate weathering tracer [16–23]. Ge is a trace element in silicate minerals, with similar geochemical properties to Si [24]. The redistribution of Ge and Si during mineral dissolution and secondary clay formation could result in variations in the Ge/Si ratio in river water and soil [17]. Generally, Ge is more enriched in secondary clays than Si [17,18,25–29]; thus, the Ge/Si ratio of river water is lower than that of soil and silicate rock [16].

For rivers on the Tibetan Plateau, the controls of the Ge/Si ratio in river water are much more complex and could be affected by multiple factors such as silicate weathering, hydrothermal water discharge, sulfide weathering, and fly coal ash [30,31]. Previous studies have reported the Ge/Si ratio in river water on the Tibetan Plateau, generally higher than the global average [23,30,31]. The elevated riverine Ge/Si ratio in the three river regions (Yangtze, Salween, and Mekong) is due to high hydrothermal activity, while in the Huang He and Hong Rivers, it may reflect the weathering of sulfide and coal-bearing minerals [23,31]. The Narayani River rises in the Great Himalaya Range in Nepal, where the Ge/Si ratio of river water is increased due to the input of the Ge-enriched hydrothermal water. Hydrothermal activities are prevailing on the southern Tibetan Plateau, forming a 2000 km long hydrothermal belt [32]. However, the proportion of hydrothermal Ge in river water and the controls and degrees of geochemical processes on the dissolved Ge/Si ratio variation at the basin scale in the southern Tibetan Plateau remains unclear.

In this study, river water and hydrothermal water samples were systematically collected in the Yarlung Tsangpo River, southern Tibetan Plateau. The Ge/Si ratios in the Yarlung Tsangpo River water show a very high level. The spatial and temporal variation of the Ge/Si ratio and the main controls of Ge/Si ratios in the Yarlung Tsangpo River water are discussed to evaluate the influence of hydrothermal input on the riverine Ge/Si ratio. Combing the literature data, the validity of the dissolved Ge/Si ratio as a silicate weathering proxy is evaluated.

2. Materials and Methods

2.1. General Settings

The Yarlung Tsangpo River is located in the southern part of Qinghai-Tibetan Plateau, with an average elevation above 4000 m. It originates from the Jemayangzong Glacier at the northern foothills of the Himalayas and traverses Tibet from west to east. At the Eastern Himalayan Syntaxis, the river changes the direction to the south, forming one of the world's largest canyons, the Yarlung Tsangpo Grand Canyon. The Yarlung Tsangpo River drains an area of 239,228 km^2, accounting for 20% of the total area and about 40.8% of the total river outflow in Tibet [33].

The strong "curtain effect" by the Himalayas prevents the wet moisture from the Indian Ocean. The upper reaches of the Yarlung Tsangpo River are subject to the plateau's semi-arid climate, and the middle reaches of the Yarlung Tsangpo River are characterized by a temperate climate. The warm and humid airflow from the Indian Ocean is transported through the lower reaches, forming a humid subtropical climate there. The rainfall mainly occurs from June to September, accounting for 65–80% of the annual precipitation [34]. Rainwater, glacial meltwater, and groundwater are the main sources of river water.

The Yarlung Tsangpo Suture Zone separates two distinct tectonic areas in the Yarlung Tsangpo River Basin, which are the Himalayan Block in the south and the Lhasa Block in the north (Figure 1). The northern boundary of the Yarlung Tsangpo Suture Zone is the Gangdise belt, a large and complex rock body consisting of granite, granodiorite, and quartz diorite. The geochemical compositions of the granite rocks in the Gangdise belt are as follows: SiO_2 (56.04–77.01%), TiO_2 (0.06–1.40%), Al_2O_3 (12.00–24.13%), FeO (0.41–7.45%), MnO (0.01–0.18%), MgO (0.08–4.69%), CaO (0.54–8.50%), Na_2O (2.28–4.79%), K_2O (0.39–6.52%), and P_2O_5 (0.01–0.27%) [35]. A 2000 km length ophiolite belt, which includes ultramafic mafic rocks, diabase, gabbro, quartz schist, marble, and exotic gneiss blocks, is located on the southern side of the Yarlung Tsangpo Suture Zone. The geochemical compositions of the ophiolitic rocks are as follows: SiO_2 (45.10–57.32%), TiO_2 (0.13–4.41%), Al_2O_3 (2.88–18.51%), FeO (1.32–13.25%), MnO (0.052–0.28%), MgO (4.58–21.66%), CaO (3.60–23.27%), Na_2O (0.10–4.22%), K_2O (0.045–7.42%), and P_2O_5 (0.006–0.879%) [36]. Due to continental collision between the South Asian Plate and the Eurasian Plate, there is an active Tethys Himalayan hydrothermal zone along the Yarlung Tsangpo Suture Zone. This hydrothermal zone possesses the most intensive hydrothermal activities in China and accounts for 80% of the hydrothermal resource in Tibet [37].

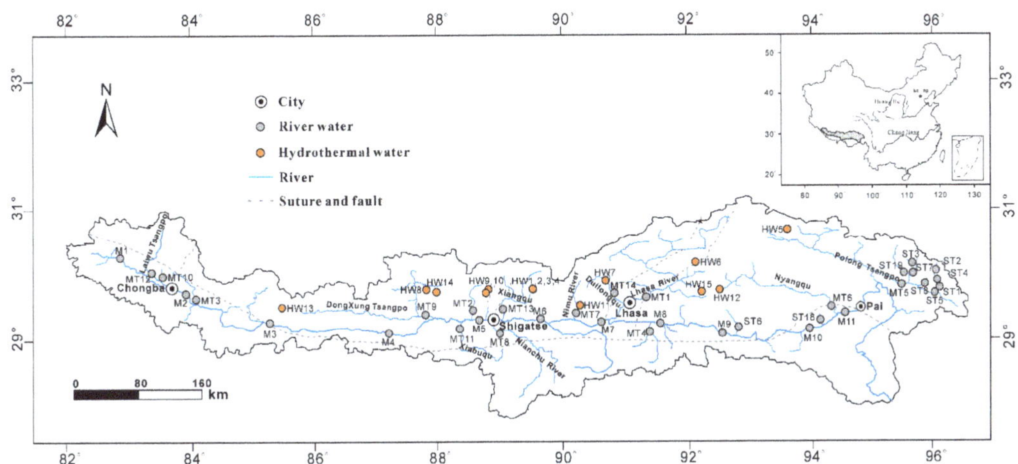

Figure 1. Sketch map of the study area and sampling sites.

2.2. Sampling and Data Measurement

The mainstream and tributary river waters and hydrothermal waters were collected in the Yarlung Tsangpo River in the high-flow period, 2018. Moreover, weekly samples were collected for the Xiangqu River, a tributary in the middle reach of the main channel. The longitude and latitude, as well as the elevation of the sampling site, were recorded by GPS (GARMIN-RINO 650). The water temperature and pH were measured in the field by a portable EC/pH meter (YSI 6600). River water samples were collected at about 5 to 10 cm below the water surface. Water samples were filtered on-site immediately with 0.22 μm Millipore cellulose acetate membranes. Filtered samples were stored in a high-density polyethylene bottle for Cl^-, SO_4^{2-} and dissolved Si concentration measurement. These containers were pre-cleaned with HNO_3 and ultrapure water (18.2 MΩ). An aliquot of samples was pre-acidified with HNO_3 to pH < 2 for Ge concentration determination.

The dissolved silicon concentration was determined using a molybdate-blue method by spectrophotometry. The measurement uncertainty of spectrophotometry was evaluated. The lab-made silicon standard solution with a known concentration was used (n = 50). After the same experimental procedure as the water sample, the measurement uncertainty was within 0.61%. Germanium concentration was measured by Perkin Elmer inductively coupled plasma mass spectrometry (model Elan DRC-e) with an uncertainty of less than 5%. The sample solution was introduced by a PFA MicroFlow nebulizer and a cyclonic spray chamber (PC3 Peltier Chiller) with a gas flow of 0.78 L min^{-1}. The readings and replicate numbers were 1 and 3, respectively. The direct analysis of trace Ge in river samples by ICP-MS is still a challenging task because major isotopes of ^{70}Ge (20.51%), ^{72}Ge (27.4%), and ^{74}Ge (36.56%) are all subject to severe polyatomic interferences. Because ^{72}Ge$^+$ was only subject to polyatomic interferences, ^{72}Ge$^+$ was chosen as the test target. By using CH_4 as the reaction gas in the dynamic reaction cell (DRC), the significant interferences, including polyatomic ions ^{36}Ar^{2+}, ^{56}Fe^{16}O$^+$, ^{40}Ar^{32}S$^+$, ^{40}Ar^{16}O^{2+}, and ^{55}Mn^{16}OH$^+$ on ^{72}Ge$^+$ determination could be successfully reduced [38]. Reproducibility (RSD) was generally less than 5% based on repeated analyses. For river water with a relatively low Ge concentration, samples were concentrated 100 times to match the detection limit. In parallel to the sample treatment procedure, all the reagent and procedural blanks were determined. Each calibration curve was evaluated by the analysis of quality control (QC) standards before, during, and after the analysis of each sample group. The chloride and sulfate concentrations were determined by Dionex Ion Chromatography (model ICS-1500) with uncertainty within 8%. All sample analysis in this study was completed in the Hydrochemistry and Environmental Laboratory at the Institute of Geology and Geophysics, Chinese Academic of Sciences.

3. Results

Data in this study are presented in Tables 1 and S1. The river water is weakly alkaline, and the pH values range from 7.25 to 9.01 (n = 45). The total dissolved solids (TDS) of river water range from 14 to 233 mg/L, Cl^- concentration of river water range from 46 to 1392 μmol/L, showing a decreasing trend from upstream to downstream. The ranges of Si and Ge concentration of river water are between 17 and 284 μmol/L and 0.1×10^{-3} and 4.6×10^{-3} μmol/L.

For hydrothermal waters, the TDS shows a large variation from 322 to >2000 mg/L, significantly higher than that of river water. The pH value of hydrothermal water ranges from 6.99 to 8.72, and the temperature varies from 43 to 90 °C. The Si concentration ranges from 598 to 2943 μmol/L, with an average value of 1399 μmol/L, which is ~13 times higher than that of river water.

Spatially, the Ge/Si ratios of river water range from 0.1 to 31.6 μmol/mol (n = 45) (Ge/Si are in 10^{-6} atom ratios hereinafter), the average value is at 7.8, an order of magnitude higher than the global average (0.6 ± 0.06, [16]). The Ge/Si value of mainstream water ranges from 2.9 to 17.1 (n = 11). The Ge/Si ratios present significant spatial variation in the Yarlung Tsangpo River Basin. In the upper and middle reaches, the river water presents a relatively high level of Ge/Si ratio (average at 13.2 and 8.8, respectively). The maximum Ge/Si value (Ge/Si = 31.60) appears in the Duilongqu (MT14), where the Yambajan Geothermal Field is located. The downstream river waters present relatively low Ge/Si ratios (average at 4.2). The lowest Ge/Si ratio is in Niequ (ST6, Ge/Si = 0.7); this might be because the river water of Niequ is mainly supplied by glacier meltwater and rainwater.

Hydrothermal waters in the Yarlung Tsangpo River Basin are enriched in Ge and Si, which span a large range of 99 to 533×10^{-3} μmol/L and 598 to 2943 μmol/L, respectively. Accordingly, there is a large range of Ge/Si ratio of 124 to 308 (average at 182) in hydrothermal water, which is 23 times higher than the average value of river water.

Table 1. Chemical compositions of river and hydrothermal water in the Yarlung Tsangpo.

Sample	Altitude (m)	pH	T (°C)	TDS (mg/L)	Ge (10^{-3} μmol/L)	Si (μmol/L)	Ge/Si (10^{-6} mol/mol)	Cl^- (μmol/L)	SO_4^{2-} (μmol/L)
					Mainstreams				
M7	3685	8.54	13	166	2.8	163	17.1	256	440
M6	3738	8.64	17	156	0.6	154	4.0	228	423
M5	3856	8.56	20	156	1.8	169	10.6	239	392
M4	4001	8.65	14	160	0.9	147	5.9	288	514
M3	4489	8.51	13	146	1.3	145	9.1	283	495
M2	4567	8.74	7	121	2.4	148	16.4	268	216
M1	4637	8.05	10	196	1.6	187	8.6	1392	303
M8	3552	8.56	18	117	1.2	144	8.5	198	371
M9	3171	8.63	17	115	0.4	131	2.9	187	377
M10	2964	8.35	19	111	0.4	106	4.0	137	312
M11	2926	7.83	15	65	0.5	94	5.1	67	198
					Main tributaries				
MT1	3621	8.54	14	135	0.8	284	2.9	200	
MT2	3890	8.32	13	105	3.3	175	18.9	127	
MT3	4570	7.96	13	143	1.5	106	13.9	419	
MT4	3558	9.01	10	233	1.0	99	9.9	172	
MT5	2726	8.41	10	66	0.2	39	3.9	65	
MT6	2982	8.22	13	25	0.2	70	2.6	55	
MT7	3694	8.11	17	107	4.6	219	21.1	170	
MT8	3807	8.39	14	170	0.7	141	4.8	200	
MT9	3943	8.18	20	112	1.8	145	12.4	235	
MT10	4595	8.02	6	105	1.5	124	11.8	216	
MT11	3903	8.14	24	115	0.6	177	3.6	209	
MT12	4587	8.60	10	111	3.1	199	15.4	151	
MT13	4300	8.51	10	158	4.5	150	29.7	343	
MT14	3745	7.25	15	129	4.6	145	31.6	606	

Table 1. *Cont.*

Sample	Altitude (m)	pH	T (°C)	TDS (mg/L)	Ge (10^{-3} μmol/L)	Si (μmol/L)	Ge/Si (10^{-6} mol/mol)	Cl$^-$ (μmol/L)	SO$_4$$^{2-}$ (μmol/L)
					Small tributaries				
ST1	3229	8.30	6	54	0.3	53	5.8	201	650
ST2	3444	8.08	10	27	0.2	42	3.6	58	44
ST3	2790	8.02	9	51	0.1	57	2.3	58	174
ST4	3390	8.20	9	56	0.3	40	8.0	64	124
ST5	3038	8.29	12	62	0.2	55	2.8	66	111
ST6	3164	7.90	10	38	0.0	17	0.7	67	124
ST7	3346	8.06	9	40	0.1	44	1.5	46	102
ST8	2788	7.85	10	33	0.1	50	2.6	57	68
ST9	3020	8.33	9	26	0.2	57	3.4	59	62
ST10	3110	8.54	10	125	0.2	68	2.6	65	878
ST11	3350	8.18	13	181	0.5	149	3.2	136	1043
ST12	3297	8.25	8	109	0.4	57	7.1	47	245
ST13	3036	7.55	9	92	0.4	67	6.1	58	728
ST14	2830	8.38	4	67	0.1	57	1.6	65	179
ST15	3032	7.90	11	75	0.3	28	10.2	66	196
ST16	2961	7.62	14	31	0.1	53	1.3	63	103
ST17	3345	7.98	9	43	0.1	25	5.6	57	91
ST18	2942	7.59	14	14	0.9	136	6.3	47	45
ST19	3303	8.07	11	54	0.1	65	1.0	47	209
					Hydrothermal water				
HW1	4581	8.72	45	>2000	102	598	170	11,099	
HW2	4596	8.68	43	>2000	99	746	133	13,797	
HW3	4580	8.62	53	1553	121	806	151	7537	
HW4	4576	7.98	64	1899	174	1212	144	4930	
HW5	4060	7.21	67	>2000	428	1387	308	26,169	
HW6	4435	7.09	44	838	142	917	155	3506	
HW7	4511	8.01	54	932	121	618	195	13,873	
HW8	4621	8.71	90	>2000	451	2180	207	15,819	
HW9	4591	7.54	71	1160	349	1880	185	3914	
HW10	4556	8.04	88	1090	331	1743	190	2994	
HW11	3898	8.63	53	322	181	1457	124	1491	
HW12	4373	8.67	52	861	203	1295	157	5314	
HW13	5010	8.62	86	900	500	2943	170	4527	
HW14	3927	7.02	65	1567	533	1840	290	15,640	
HW15	4275	6.99	49	1124	207	1370	151	5669	

4. Discussion

4.1. Different Sources of Ge Affecting Ge/Si Ratios in the River

Dissolved Ge in the river is generally derived from fly ash generated from coal combustion, weathering of silicate minerals and sulfide, and hydrothermal water. Coal is an important raw material for Ge [39]. Germanium is easily hydrolyzed and generally exists in the form of Ge(OH)$_4$ in fluid, thus preferentially enriched in river water compared with other trace elements such as Cu and Zn derived from fly ash [31]. Rivers that drain industrial areas present relatively high Ge/Si ratios resulting from coal combustion [40]. However, the anthropogenic source is considered to be negligible in the Yarlung Tsangpo River Basin because the economic structure here is mainly composed of agriculture, animal husbandry, and tourism. A large number of residents still use straw and air-dried cow dung as their energy source [41].

Silicate weathering dominates the dissolved Ge and Si in large rivers such as the Mackenzie River [23] and the Amazon River [42]. The formation of secondary minerals during silicate weathering would lead to redistribution of Ge and Si in soil and river water [26]. Since Ge preferentially incorporates into secondary minerals compared with Si, the Ge/Si ratio of river water would be lower than that of bedrock [20,25]. In addition, the weathering of plagioclase (Ge/Si of ~1.5) to kaolinite (Ge/Si of 4.8–6.1) results in a decrease in the riverine Ge/Si ratio, and kaolinite weathering will lead to an increase in dissolved Ge/Si ratio in the solution [19,20,25]. In the Yarlung Tsangpo River, the water shows an extremely high Ge/Si ratio (average at 7.8), which is significantly higher than that of plagioclase and kaolinite.

Sulfide weathering has been considered as a possible process affecting the Ge/Si ratio in many large rivers such as the Copper River [18], the Red River [7], and the Yellow River [9,31]. In the Yarlung Tsangpo River, there is a strong relationship between dissolved Cl^- and SO_4^{2-} in the mainstream (except for samples in the source areas M1 and M2, Figure 2), which suggests that SO_4^{2-} is mainly derived from hydrothermal spring or evaporite dissolution, rather than sulfide mineral oxidation.

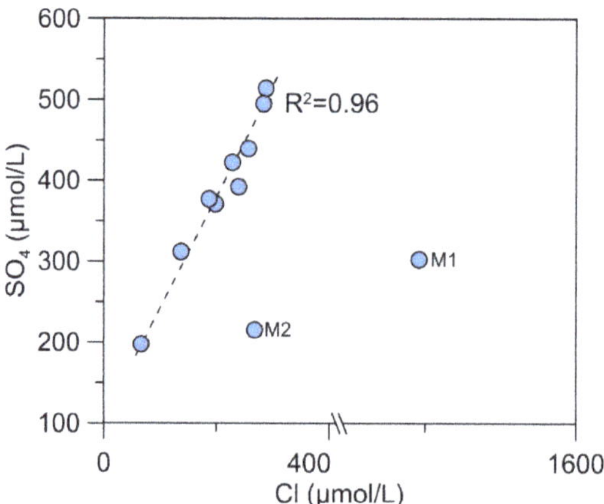

Figure 2. The relationship between Cl^- and SO_4^{2-} in mainstream waters.

The hydrothermal water samples collected in this river basin have extremely high Si and Ge concentrations and Ge/Si ratios, which might be the result of water–rock interactions during the upwelling of hydrothermal fluids. As shown in Figure 3, Ge, Si, and the temperature of hydrothermal water are positively correlated. Germanium is more enriched than silicon due to the precipitation of quartz during the cooling process of hydrothermal water. Therefore, temperature is the vital factor affecting Ge and Si in hydrothermal waters [43,44].

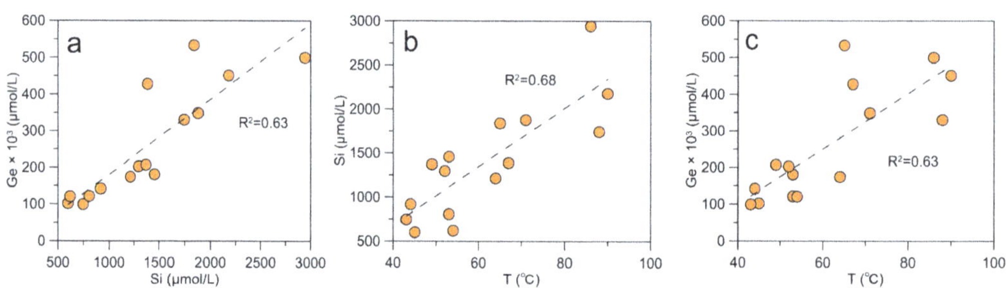

Figure 3. Correlations between dissolved Si and Ge (**a**), water temperature and Si (**b**), and water temperature and Ge (**c**) in hydrothermal waters.

4.2. Ge/Si Ratios in the Yarlung Tsangpo River

4.2.1. Spatial and Temporal Variations of Riverine Ge/Si Ratio

The Ge/Si ratios of river water in the Yarlung Tsangpo show large variations, roughly decrease from upstream to downstream, and show a positive relation with elevation (Figure 4). The correlations between Ge and Si concentrations and the TDS and Ge/Si ratio of river water suggest that the Ge and Si contents of some river waters are significantly affected by the contribution of hydrothermal water (Figure 5), such as samples MT7, MT13, and MT14, with the highest Ge/Si ratio, the corresponding river flow through the fault zone where hot springs are widely distributed. The river waters with a low Ge/Si ratio are generally small tributaries in the lower reaches of the Yarlung Tsangpo (Figure 4a). Even so, the Ge/Si ratios of these small tributaries (0.1–10.2, average at 3.8) are still far above the world average. One possible explanation is hydrothermal water input. However, the Cl^- content in these rivers is relatively low, similar to river ST15, the Ge/Si ratio can reach up to 10.2, while the Cl^- concentration is 66 µmol/L, significantly lower than that in mainstreams, and thus the influence of the input of hydrothermal water is limited. Another potential reason accounting for the relatively high Ge/Si ratios could be sulfide weathering because of the high sulfate concentration (44–1043 µmol/L, average at 276 µmol/L) of river waters. The terrain in these river basins is steep, with some glacier-covered, and erosion is intense; this promotes the sulfide weathering process. As shown in Figure 6, the Ge/Si ratios of the mainstream and the main tributaries of the Yarlung Tsangpo River are affected by hydrothermal water input, while the controls of the Ge/Si ratio of small tributaries downstream are complicated. Besides sulfide weathering, weathering of minerals such as biotite could also affect the dissolved Ge signals in the river water under glacier cover. In Figure 6, there are four rivers located below the global river line, and a strong correlation ($R^2 = 0.98$) between Ge/Si ratios and K^+/Na^+ ratios are observed in these four samples. Considering biotite is rich in K and Ge (Ge content is 3–4 times higher than the upper continental crust) [25,45], and glacial erosion could lead to exposure of fresh minerals such as biotite, one of the most vulnerable silicate minerals to weathering, though accounting for a small proportion of silicate rocks [18].

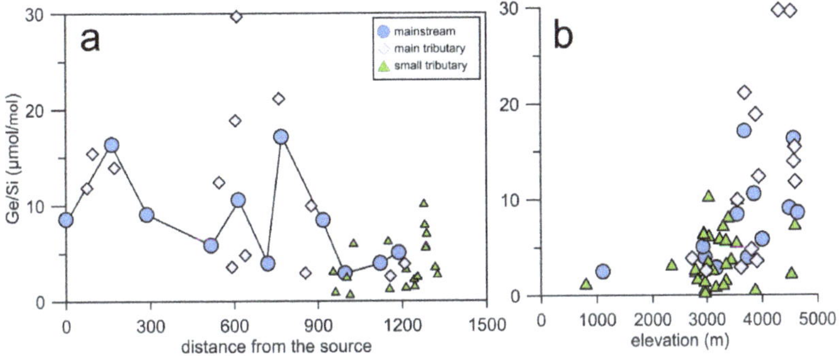

Figure 4. Spatial variations in Ge/Si ratio of the Yarlung Tsangpo River. (**a**) Relations between Ge/Si ratio and distance from the source; (**b**) Relations between Ge/Si ratio and elevation.

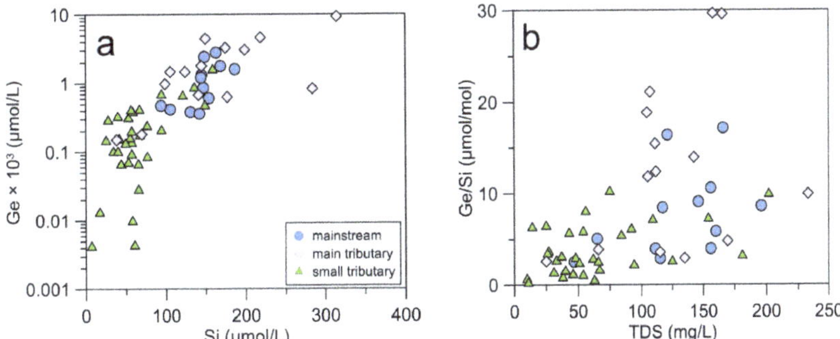

Figure 5. Relations between Si and Ge (**a**) and TDS and Ge/Si ratio (**b**) of river water.

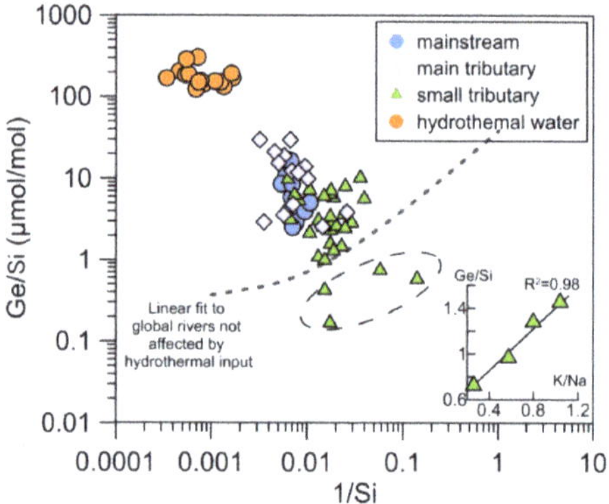

Figure 6. The relation between Si and Ge/Si in river water. The line for global clean rivers is from [17]. Adapted with permission from ref [17]. Copyright 1992 John Wiley and Sons.

The tributary Xiangqu is characterized by the distribution of geothermal springs around its source area. The Ge/Si ratio of weekly time series river water samples in the Xiangqu show an obvious relationship with river discharge. As discharge increases, the Ge/Si ratio of river water decreases (Figure 7). In rivers affected by hydrothermal inputs, such as the mainstream of the Yarlung Tsangpo River, Xiangqu, and Nepali rivers [30], a low Ge/Si ratio of river water is observed at the high-flow period. This suggests the dilution of the hydrothermal water input during the high-flow period. However, in the silicate weathering controlled system, the trend is the opposite. The Ge/Si of river water is generally positively related to river water discharge (Figure 8). The water–rock reaction may be relatively strong in the whole basin, and the clay weathering process, which can produce high Ge/Si ratio fluids, is more obvious during the high-flow season. This is consistent with previous observations [16,20,21].

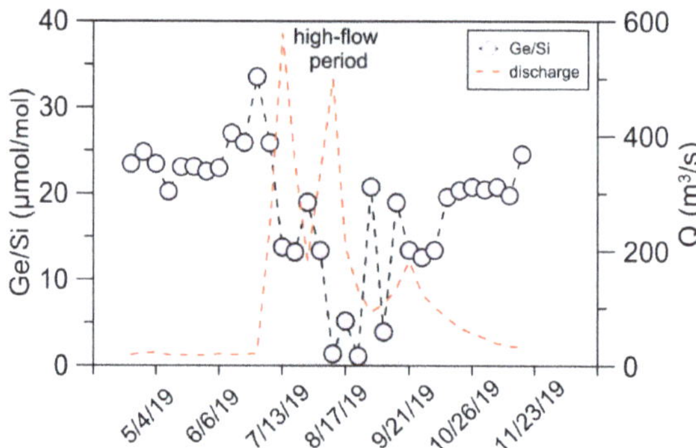

Figure 7. Ge/Si ratio variation in time series river water samples in Xiangqu.

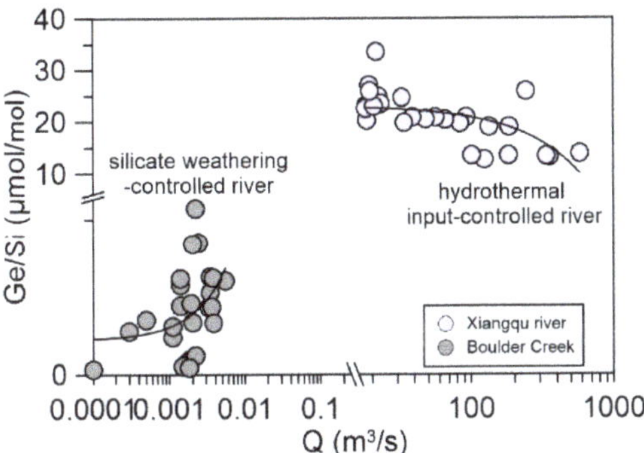

Figure 8. The plot of discharge versus Ge/Si ratio in silicate weathering- and hydrothermal input-controlled rivers. Data of Boulder Creek streams is from [46]. Adapted with permission from ref [46]. Copyright 2017 John Wiley and Sons.

4.2.2. Ge and Si in Rivers Supplied by Hydrothermal Water

Chloride was considered to be conservative during the mixing process of hydrothermal spring and river water and, thus, supposed as an effective tracer of hydrothermal input of Ge and Si in the Yarlung Tsangpo River [43]. Assuming the Cl^- concentration of river water is the weighted sum of rainfall and hydrothermal contribution, a mean value of rainwater in southeastern Tibet is 6.7 µmol/L [47], and an evapotranspiration factor of 2, the Cl^- originating from rainfall is 13.4 µmol/L. Riverine Ge and Si derived from hydrothermal water can be calculated according to the following equations:

$$[Cl]_{hydr} = [Cl]_{riv} - [Cl]_{atm} \tag{1}$$

$$[Ge]_{hydr} = [Cl]_{hydr} \times (Ge/Cl)_{hydr} \tag{2}$$

$$[Si]_{hydr} = [Cl]_{hydr} \times (Si/Cl)_{hydr} \tag{3}$$

where $(Ge/Cl)_{hydr}$ and $(Si/Cl)_{hydr}$ represent the average ratio of hydrothermal waters in each sub-basin in the Yarlung Tsangpo River. $[Ge]_{hydr}$ and $[Si]_{hydr}$ represent riverine Ge and Si derived from hydrothermal waters. Potential uncertainties for calculating $[Ge]_{hydr}$ and $[Si]_{hydr}$ are dominantly from the value of $(Ge/Cl)_{hydr}$ and $(Si/Cl)_{hydr}$ in sub-basins due to the limited numbers of hydrothermal samples. The Xiangqu, Nimu River, Lhasa River, and Duilongqu were selected in this calculation because hydrothermal springs were collected in these basins. We calculated that 4–98% of Si is from hydrothermal water (Table S2). The calculation also shows an unexpected result that $[Ge]_{hydr}$ is much higher than dissolved Ge in the river water. This disparity suggests that a significant amount of Ge has been lost during its transportation from spring to river. A potential Ge-enriched component is clays [16,25,30]. Moreover, hydrothermal water could contain high contents of sulfide and iron—when it comes to the surface, the sulfur is deposited as sulfate, and iron deposited as Fe-oxyhydroxide, which are also enriched in Ge [39]. Thus, there should be two possible explanations for significant Ge depletion: absorbed on clays in soils and sediments and deposited in sulfate and Fe-oxyhydroxides. It is also estimated that at least 11–88% of Ge incorporates into the solid phase in these river basins (Table S2). This is similar to the result that 70–90% of Ge in seafloor hydrothermal water hosts in solid during Fe-oxyhydroxides precipitation [48]. Such a proportion of Ge incorporated into solid is much more than clay sequestration during silicate weathering [28]. Nevertheless, the Ge/Si ratio of rivers on the Tibetan Plateau is ~10 times higher than that of the global average; thus, the Tibetan Plateau tectonism during the Cenozoic, which could enhance hydrothermal activity, might increase the Ge/Si ratio of the continental rivers and seawater.

4.3. Using Ge/Si Ratio to Trace Geochemical Processes

The Ge/Si ratios of large rivers in the different climatic and geological backgrounds are mainly controlled by weathering processes, hydrothermal input, and anthropogenic activity such as coal combustion [25,31,42]. Compared with other large rivers in the world, the Yarlung Tsangpo River waters present the highest average Ge/Si ratio.

As Figure 9 shows, the average value of Ge/Si ratios in rivers such as the Amazon River, Mackenzie River, Yukong River, and the Orinoco River are close to the world average (Ge/Si = 0.6 ± 0.06). These rivers drain regions without obvious input from hydrothermal and industrial activity, and the Ge/Si ratios are mainly controlled by silicate weathering processes [7,9,16,23,42]. River water in the eastern Qinghai-Tibet Plateau (Salween, Mekong, and the upstream of Yangtze) show relatively high Ge/Si values (average at ~5.59, [31]). It is similar to Ge/Si in the Yarlung Tsangpo River, where hydrothermal water discharge dominates the riverine Ge/Si ratio.

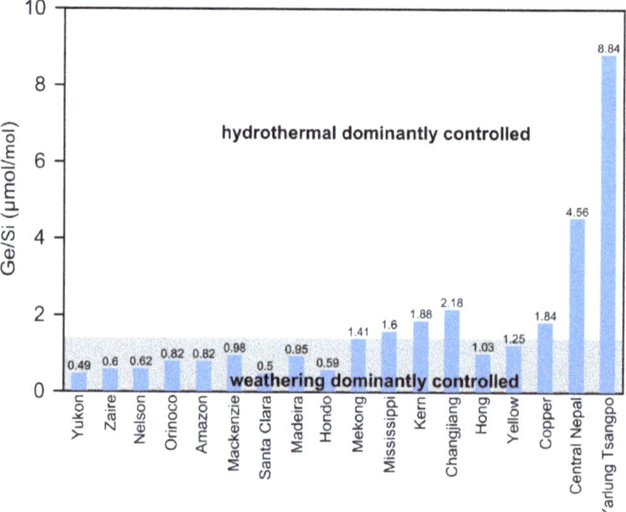

Figure 9. Ge/Si ratio histogram of the Yarlung Tsangpo River and other typical basins. Data of Zaire River, Yukong River, Orinoco River, and Nelson River are from [16]. Data of Amazon River, Mackenzie River, Mekong River, Mississippi River, Changjiang River, Kern River, Santa Clara River, Madeira River, and Hondo River are from [23]. Data of Hong River is from [7]. Data on the Yellow River is from [9]. Data of Copper River is from [18]. Data of Central Nepal Rivers is from [30].

Compiling these results, a simplified category can be made for riverine Ge/Si ratios (Figure 9). Two characteristic Ge/Si values were chosen as the division standard, which is 1.5 (average value of primary silicate rock) and 3.75 (average Ge/Si ratio of secondary aluminosilicate clays) [25,40]. Generally, for river water, the Ge/Si ratio is lower than ~1.5, and the Ge/Si ratios of river water reflect the degree of silicate weathering. For rivers, the Ge/Si ratio is higher than ~3.75, and the hydrothermal effect should be regarded as the dominant control. Additionally, the impact of sulfide weathering on the dissolved Ge/Si ratios should be considered for a basin with a relatively high sulfate concentration. The controlling factors of river waters in this range need to be discussed in terms of the distribution of hydrothermal waters and sulfide minerals weathering. In addition, the controlling factors of dissolved Ge/Si ratio are complicated in rivers with Ge/Si ratios ranging from ~1.5 to ~3.75. It might be the combined effect of silicate weathering and hydrothermal input and possibly involve anthropogenic activities. For instance, Mississippi River water is apparently affected by urban and industrial input, which increased the Ge concentration of river water to 266×10^{-3} μmol/L, and the Ge/Si ratio is 1.60 [23]. Changjiang River water is influenced by extensive irrigation and rice cultivation, which could cause a decrease in Si concentration in the river water; the Ge/Si ratio in the Changjiang River is 2.18 [49].

5. Conclusions

The spatial and temporal variations of Ge/Si ratios indicate a significant hydrothermal water impact on the riverine Ge/Si ratios in the river basin. According to the mass balance equation, the calculated riverine Ge sourced from hydrothermal waters is much higher than measured Ge in the river water. This indicates that a large amount of Ge has been depleted in the river system. The absorption of Ge on clays or deposition in sulfate and Fe-oxyhydroxide minerals could be a plausible explanation for such significant Ge depletion. The Ge/Si ratios of the Yarlung Tsangpo River water evidently reflect the discharge of hydrothermal water, which indicates that the hydrothermal water contribution should be carefully considered when using the Ge/Si ratio to trace silicate weathering in the Yarlung Tsangpo River Basin.

Supplementary Materials: The following are available online at https://www.mdpi.com/article/10.3390/w14020181/s1, Table S1: time series variations in Ge/Si ratio of river water in Xiangqu and the mainstream of the Yarlung Tsangpo River, Table S2: calculation of Ge_{hydr}, Si_{hydr}, and Ge_{lost}.

Author Contributions: Conceptualization, Y.W. and Z.X.; methodology, Y.W., T.Z., Z.X. and J.Z.; software, Y.W.; validation, Y.W., T.Z. and Z.X.; formal analysis, Y.W. and T.Z.; investigation, Y.W., T.Z., Z.X. and H.S.; resources, Z.X.; data curation, Y.W., T.Z., Z.X., H.S. and J.Z.; writing—original draft preparation, Y.W., T.Z. and Z.X.; writing—review and editing, Y.W., T.Z., Z.X., H.S. and J.Z.; visualization, Y.W., T.Z., Z.X., H.S. and J.Z.; supervision, Z.X.; project administration, Z.X.; funding acquisition, Z.X. All authors have read and agreed to the published version of the manuscript.

Funding: This research was financially supported by the National Natural Science Foundation of China (Grant Nos. 91747202, 41730857, 42003015), the "Strategic Priority Research Program" of the Chinese Academy of Sciences (Grant No. XDB26000000) and the National Key Research and Development Program of China (Grant No. 2020YFA0607700).

Institutional Review Board Statement: Not applicable.

Informed Consent Statement: Not applicable.

Data Availability Statement: The data presented in this study are available on request from the corresponding author.

Conflicts of Interest: The authors declare no conflict of interest.

References

1. Berner, R.A.; Lasaga, A.C.; Garrels, R.M. The carbonate-silicate geochemical cycle and its effect on atmospheric carbon dioxide over the past 100 million years. *Am. J. Sci.* **1983**, *283*, 641–683. [CrossRef]
2. Walker, J.C.G.; Hays, P.B.; Kasting, J.F. A negative feedback mechanism for the long-term stabilization of Earth's surface temperature. *J. Geophys. Res. Ocean.* **1981**, *86*, 9776–9782. [CrossRef]
3. Raymo, M.E.; Ruddiman, W.F. Tectonic forcing of late Cenozoic climate. *Nature* **1992**, *359*, 117–122. [CrossRef]
4. France-Lanord, C.; Derry, L.A. Organic carbon burial forcing of the carbon cycle from Himalayan erosion. *Nature* **1997**, *390*, 65–67. [CrossRef]
5. Gaillardet, J.; Dupré, B.; Louvat, P.; Allègre, C.J. Global silicate weathering and CO_2 consumption rates deduced from the chemistry of large rivers. *Chem. Geol.* **1999**, *159*, 3–30. [CrossRef]
6. Hren, M.T.; Chamberlain, C.P.; Hilley, G.E.; Blisniuk, P.M.; Bookhagen, B. Major ion chemistry of the Yarlung Tsangpo–Brahmaputra river: Chemical weathering, erosion, and CO_2 consumption in the southern Tibetan plateau and eastern syntaxis of the Himalaya. *Geochim. Cosmochim. Acta* **2007**, *71*, 2907–2935. [CrossRef]
7. Moon, S.; Huh, Y.; Qin, J.; van Pho, N. Chemical weathering in the Hong (Red) River basin: Rates of silicate weathering and their controlling factors. *Geochim. Cosmochim. Acta* **2007**, *71*, 1411–1430. [CrossRef]
8. Noh, H.; Huh, Y.; Qin, J.; Ellis, A. Chemical weathering in the Three Rivers region of Eastern Tibet. *Geochim. Cosmochim. Acta* **2009**, *73*, 1857–1877. [CrossRef]
9. Wu, L.; Huh, Y.; Qin, J.; Du, G.; van Der Lee, S. Chemical weathering in the Upper Huang He (Yellow River) draining the eastern Qinghai-Tibet Plateau. *Geochim. Cosmochim. Acta* **2005**, *69*, 5279–5294. [CrossRef]

10. Huh, Y. Chemical weathering and climate—A global experiment: A review. *Geosci. J.* **2003**, *7*, 277–288. [CrossRef]
11. Struyf, E.; Smis, A.; Van Damme, S.; Meire, P.; Conley, D.J. The Global Biogeochemical Silicon Cycle. *Silicon* **2009**, *1*, 207–213. [CrossRef]
12. Han, G.; Liu, C.-Q. Water geochemistry controlled by carbonate dissolution: A study of the river waters draining karst-dominated terrain, Guizhou Province, China. *Chem. Geol.* **2004**, *204*, 1–21. [CrossRef]
13. El Mrissani, S.; Haida, S.; Probst, J.-L.; Probst, A. Multi-Indices Assessment of Origin and Controlling Factors of Trace Metals in River Sediments from a Semi-Arid Carbonated Basin (the Sebou Basin, Morocco). *Water* **2021**, *13*, 3203. [CrossRef]
14. Alexakis, D. Diagnosis of stream sediment quality and assessment of toxic element contamination sources in East Attica, Greece. *Environ. Earth Sci.* **2011**, *63*, 1369–1383. [CrossRef]
15. Papadopoulou-Vrynioti, K.; Alexakis, D.; Bathrellos, G.D.; Skilodimou, H.D.; Vryniotis, D.; Vassiliades, E.; Gamvroula, D. Distribution of trace elements in stream sediments of Arta plain (western Hellas): The influence of geomorphological parameters. *J. Geochem. Explor.* **2013**, *134*, 17–26. [CrossRef]
16. Mortlock, R.A.; Frohlich, P.N. Continental weathering of germanium: GeSi in the global river discharge. *Geochim. Cosmochim. Acta* **1987**, *51*, 2075–2082. [CrossRef]
17. Froelich, P.N.; Blanc, V.; Mortlock, R.A.; Chillrud, S.N.; Dunstan, W.; Udomkit, A.; Peng, T.-H. River Fluxes of Dissolved Silica to the Ocean Were Higher during Glacials: Ge/Si In Diatoms, Rivers, and Oceans. *Paleoceanogr. Paleoclimatol.* **1992**, *7*, 739–767. [CrossRef]
18. Anders, A.M.; Sletten, R.S.; Derry, L.A.; Hallet, B. Germanium/silicon ratios in the Copper River Basin, Alaska: Weathering and partitioning in periglacial versus glacial environments. *J. Geophys. Res. Earth. Surf.* **2003**, *108*. [CrossRef]
19. Derry, L.A.; Pett-Ridge, J.C.; Kurtz, A.C.; Troester, J.W. Ge/Si and 87Sr/86Sr tracers of weathering reactions and hydrologic pathways in a tropical granitoid system. *J. Geochem. Explor.* **2006**, *88*, 271–274. [CrossRef]
20. Lugolobi, F.; Kurtz, A.C.; Derry, L.A. Germanium–silicon fractionation in a tropical, granitic weathering environment. *Geochim. Cosmochim. Acta* **2010**, *74*, 1294–1308. [CrossRef]
21. Kurtz, A.C.; Lugolobi, F.; Salvucci, G. Germanium-silicon as a flow path tracer: Application to the Rio Icacos watershed. *Water Resour. Res.* **2011**, *47*, W06516. [CrossRef]
22. Meek, K.; Derry, L.; Sparks, J.; Cathles, L. 87Sr/86Sr, Ca/Sr, and Ge/Si ratios as tracers of solute sources and biogeochemical cycling at a temperate forested shale catchment, central Pennsylvania, USA. *Chem. Geol.* **2016**, *445*, 84–102. [CrossRef]
23. Baronas, J.J.; Torres, M.A.; West, A.J.; Rouxel, O.; Georg, B.; Bouchez, J.; Gaillardet, J.; Hammond, D.E. Ge and Si isotope signatures in rivers: A quantitative multi-proxy approach. *Earth Planet. Sci. Lett.* **2018**, *503*, 194–215. [CrossRef]
24. Goldschmidt, V.M. Concerning the crystallo-chemical and geochemical behaviour of Germanium. *Naturwissenschaften* **1926**, *14*, 295–297. [CrossRef]
25. Kurtz, A.C.; Derry, L.A.; Chadwick, O.A. Germanium-silicon fractionation in the weathering environment. *Geochim. Cosmochim. Acta* **2002**, *66*, 1525–1537. [CrossRef]
26. Scribner, A.M.; Kurtz, A.C.; Chadwick, O.A. Germanium sequestration by soil: Targeting the roles of secondary clays and Fe-oxyhydroxides. *Earth Planet. Sci. Lett.* **2006**, *243*, 760–770. [CrossRef]
27. Opfergelt, S.; Cardinal, D.; André, L.; Delvigne, C.; Bremond, L.; Delvaux, B. Variations of δ30Si and Ge/Si with weathering and biogenic input in tropical basaltic ash soils under monoculture. *Geochim. Cosmochim. Acta* **2010**, *74*, 225–240. [CrossRef]
28. Qi, H.-W.; Hu, R.-Z.; Jiang, K.; Zhou, T.; Liu, Y.-F.; Xiong, Y.-W. Germanium isotopes and Ge/Si fractionation under extreme tropical weathering of basalts from the Hainan Island, South China. *Geochim. Cosmochim. Acta* **2019**, *253*, 249–266. [CrossRef]
29. Perez-Fodich, A.; Derry, L.A. A model for germanium-silicon equilibrium fractionation in kaolinite. *Geochim. Cosmochim. Acta* **2020**, *288*, 199–213. [CrossRef]
30. Evans, M.J.; Derry, L.A.; France-Lanord, C. Geothermal fluxes of alkalinity in the Narayani river system of central Nepal. *Geochem. Geophys. Geosyst.* **2004**, *5*, Q08011. [CrossRef]
31. Han, Y.; Huh, Y.; Derry, L. Ge/Si ratios indicating hydrothermal and sulfide weathering input to rivers of the Eastern Tibetan Plateau and Mt. Baekdu. *Chem. Geol.* **2015**, *410*, 40–52. [CrossRef]
32. Zhang, M.; Guo, Z.; Zhang, L.; Sun, Y.; Cheng, Z. Geochemical constraints on origin of hydrothermal volatiles from southern Tibet and the Himalayas: Understanding the degassing systems in the India-Asia continental subduction zone. *Chem. Geol.* **2017**, *469*, 19–33. [CrossRef]
33. Liu, T.-C. Hydrological features of the Yarlung Tsangpo River. *Acta Geogr. Sin.* **1999**, *54*, 157–164, (In Chinese with English Abstract).
34. Huang, X.; Sillanpää, M.; Gjessing, E.T.; Peräniemi, S.; Vogt, R.D. Water quality in the southern Tibetan Plateau: Chemical evaluation of the Yarlung Tsangpo (Brahmaputra). *River Res. Appl.* **2011**, *27*, 113–121. [CrossRef]
35. Chen, T. Geochemistry of the Qushui Intrusive of Gangdese in Tibet and Its Implications for Magma Mixing. Master's Thesis, China University of Geosciences, Beijing, China, 2006. (In Chinese).
36. Geng, Q.; Pan, G.; Zheng, L.; Sun, Z.; Ou, C.; Dong, H. Petrological characteristics and original settings of the Yarlung Tsangpo ophiolitic mélange in Namche Barwa, SE Tibet. *Sci. Geol. Sin.* **2004**, *39*, 388–406.
37. Tan, H.; Zhang, Y.; Zhang, W.; Kong, N.; Zhang, Q.; Huang, J. Understanding the circulation of geothermal waters in the Tibetan Plateau using oxygen and hydrogen stable isotopes. *Appl. Geochem.* **2014**, *51*, 23–32. [CrossRef]

38. Zhang, J.; Wang, X.; Dong, Y.; Xu, Z. Direct determination of trace germanium in river samples by CH4-Ar mixed gas plasma DRC-ICP-MS. *Int. J. Mass Spectrom.* **2017**, *421*, 184–188. [CrossRef]

39. Bernstein, L.R. Germanium geochemistry and mineralogy. *Geochim. Cosmochim. Acta* **1985**, *49*, 2409–2422. [CrossRef]

40. Froelich, P.N.; Hambrick, G.A.; Andreae, M.O.; Mortlock, R.A.; Edmond, J.M. The geochemistry of inorganic germanium in natural waters. *J. Geophys. Res. Ocean.* **1985**, *90*, 1133–1141. [CrossRef]

41. Guo, C.; Peng, O. On the Relationship between the Specialty Set-Up of Secondary Vocational Education and Regional Industrial Structure in Tibet. *Can. Soc. Sci.* **2015**, *11*, 244–248.

42. Baronas, J.J.; Hammond, D.E.; McManus, J.; Wheat, C.G.; Siebert, C. A global Ge isotope budget. *Geochim. Cosmochim. Acta* **2017**, *203*, 265–283. [CrossRef]

43. Evans, M.J.; Derry, L.A. Quartz control of high germanium/silicon ratios in geothermal waters. *Geology* **2002**, *30*, 1019–1022. [CrossRef]

44. Pokrovski, G.S.; Roux, J.; Hazemann, J.-L.; Testemale, D. An X-ray absorption spectroscopy study of argutite solubility and aqueous Ge(IV) speciation in hydrothermal fluids to 500 °C and 400 bar. *Chem. Geol.* **2005**, *217*, 127–145. [CrossRef]

45. Burton, J.D.; Culkin, F.; Riley, J.P. The abundances of gallium and germanium in terrestrial materials. *Geochim. Cosmochim. Acta* **1959**, *16*, 151–180. [CrossRef]

46. Aguirre, A.A.; Derry, L.A.; Mills, T.J.; Anderson, S.P. Colloidal transport in the Gordon Gulch catchment of the Boulder Creek CZO and its effect on C-Q relationships for silicon. *Water Resour. Res.* **2017**, *53*, 2368–2383. [CrossRef]

47. Liu, B.; Kang, S.; Sun, J.; Zhang, Y.; Xu, R.; Wang, Y.; Liu, Y.; Cong, Z. Wet precipitation chemistry at a high-altitude site (3326 m a.s.l.) in the southeastern Tibetan Plateau. *Environ. Sci. Pollut. Res.* **2013**, *20*, 5013–5027. [CrossRef] [PubMed]

48. Escoube, R.; Rouxel, O.J.; Edwards, K.; Glazer, B.; Donard, O.F.X. Coupled Ge/Si and Ge isotope ratios as geochemical tracers of seafloor hydrothermal systems: Case studies at Loihi Seamount and East Pacific Rise 9°50′N. *Geochim. Cosmochim. Acta* **2015**, *167*, 93–112. [CrossRef]

49. Ding, T.; Wan, D.; Wang, C.; Zhang, F. Silicon isotope compositions of dissolved silicon and suspended matter in the Yangtze River, China. *Geochim. Cosmochim. Acta* **2004**, *68*, 205–216. [CrossRef]

Article

Geochemistry of Dissolved Heavy Metals in Upper Reaches of the Three Gorges Reservoir of Yangtze River Watershed during the Flood Season

Jie Zeng [1], Guilin Han [1,*], Mingming Hu [2,3], Yuchun Wang [2,3], Jinke Liu [1], Shitong Zhang [1] and Di Wang [1]

[1]   Institute of Earth Sciences, China University of Geosciences (Beijing), Beijing 100083, China; zengjie@cugb.edu.cn (J.Z.); jinkeliu@cugb.edu.cn (J.L.); stongzhang0103@cugb.edu.cn (S.Z.); wangdi1123@cugb.edu.cn (D.W.)

[2]   State Key Laboratory of Simulation and Regulation of Water Cycle in River Basin, Beijing 100083, China; humingming@iwhr.com (M.H.); wangyc@iwhr.com (Y.W.)

[3]   Department of Water Ecology and Environment, China Institute of Water Resources and Hydropower Research, Beijing 100083, China

*   Correspondence: hanguilin@cugb.edu.cn; Tel.: +86-10-8232-3536

Abstract: Dissolved heavy metals (HMs), derived from natural and anthropogenic sources, are an important part of aquatic environment research and gain more international concern due to their acute toxicity. In this study, the geochemistry of dissolved HMs was analyzed in the upper Three Gorges Reservoir (TGR) of the Yangtze River (YZR) watershed to explore their distribution, status, and sources and further evaluate the water quality and HM-related risks. In total, 57 water samples were collected from the main channel and tributaries of the upper TGR. The concentrations of eight HMs, namely V, Ni, Cu, Zn, As, Mo, Cd, and Pb, were measured by ICP-MS. The mean concentrations (in μg/L) of eight HMs decreased in the order: As (1.46), V (1.44), Ni (1.40), Mo (0.94), Cu (0.86), Zn (0.63), Pb (0.03), and Cd (0.01). The concentrations of most HMs were 1.4~8.1 times higher than that in the source area of the YZR, indicating a potential anthropogenic intervention in the upper TGR. Spatially, the concentrations of V, Cu, As, and Pb along the main channel gradually decreased, while the others were relatively stable (except for Cd). The different degrees of variations in HM concentrations were also found in tributaries. According to the correlation analysis and principal component (PC) analysis, three PCs were identified and explained 75.1% of the total variances. combined with the concentrations of each metal, PC1 with high loadings of V, Ni, As, and Mo was considered as the main contribution of human inputs, PC2 (Cu and Pb) was primarily attributed to the contribution of mixed sources of human emissions and natural processes, and Zn and Cd in PC3 were controlled by natural sources. Water quality assessment suggested the good water quality (meeting the requirements for drinking purposes) with WQI values of 14.1 ± 3.4 and 11.6 ± 3.6 in the main channel and tributaries, respectively. Exposure risk assessment denoted that the health effects of selected HMs on the human body were limited (hazard index, HI < 1), but the potential risks of V and As with HI > 0.1 were non-negligible, especially for children. These findings provide scientific support for the environmental management of the upper TGR region and the metal cycle in aquatic systems.

Keywords: heavy metals; river pollution; source identification; risk evaluation; upper Yangtze river watershed

Citation: Zeng, J.; Han, G.; Hu, M.; Wang, Y.; Liu, J.; Zhang, S.; Wang, D. Geochemistry of Dissolved Heavy Metals in Upper Reaches of the Three Gorges Reservoir of Yangtze River Watershed during the Flood Season. _Water_ **2021**, _13_, 2078. https:// doi.org/10.3390/w13152078

Academic Editor: Laura Bulgariu

Received: 18 July 2021
Accepted: 29 July 2021
Published: 30 July 2021

Publisher's Note: MDPI stays neutral with regard to jurisdictional claims in published maps and institutional affiliations.

1. Introduction

With the acceleration of economic globalization, it becomes more and more difficult to balance the excessive utilization of water resources and high-efficiency water environmental protection schemes [1,2]. Given the current status of the global hydrosphere, it is of great significance to evaluate the contamination of river ecosystems (watershed scale), which is beneficial for the arrangement of water resources [3–5]. Heavy metal (HM) pollution is one of the most common contamination problems in aquatic environments [6,7]. Characterized by bioaccumulation, non-biodegradability, and acute/chronic toxicity [8–12], HMs are harmful to the aquatic ecological environment and even threaten human health [13]. Both anthropogenic inputs (e.g., industrial/domestic wastes) and natural processes (e.g., rock weathering and soil erosion) can contribute to the riverine HMs [14,15]. Moreover, the occurrences of HMs in water bodies include dissolved loads, riverbed sedimentary loads, and suspended loads [16].

Among these three occurrences, dissolved HMs are considered to be more harmful to humans or aquatic organisms due to more potential exposure pathways [17,18]. The two main exposure routes of HMs are direct ingestion (drinking water) and dermal absorption (via skin) [19,20]. According to previous studies, chemical carcinogens resulted in 90% of cancers, and the ingestion of drinking water is a significant route of these chemicals [21]. Obviously, HM contamination in rivers has a high correlation with the human health risk. In addition, low doses of HMs can also be detrimental to the human body during long-term exposure [22]. Therefore, numerous watershed-scale studies have been carried out on dissolved HMs in rivers all over the world. The results show that there is high heterogeneity with regard to the state of heavy metal compositions and their environmental/risk effects in the aquatic environment [20,23–26]. Under the background of anthropogenic disturbances, the river becomes quite sensitive to pollution at the watershed scale, which makes the river a reflection of the impact of both human disturbance and natural processes, and it further reflects the potential risks or negative effects on life [27]. Beside this, the re-released processes of HMs from the suspended/bed sediments are also significant if the ambient conditions (e.g., pH) change. Therefore, understanding and identifying the contamination, sources, and risks of HMs in river water is important and beneficial for efficient and sustainable water resources management.

As the largest watershed in Asia, the Yangtze River watershed (YRW) originates from the Tibetan Plateau, breeds a variety of ecosystems, and provides freshwater resources for hundreds of millions of people. However, the YRW is highly sensitive to climate change and human perturbations (pollution) [21,28]. Due to the abundant river flow and appropriate natural elevation drop, YRW is an ideal place for hydropower generation. Therefore, more than 50,000 reservoirs/dams distribute throughout the YRW, such as the Three Gorges Reservoir (TGR) and Gezhouba Dam [29,30], which powerfully support the local economy. Reservoir development and other human activities (e.g., urban emission and agricultural production), have significantly changed hydrodynamic conditions (e.g., hydraulic residence time, HRT) and further affected river environmental processes and biogeochemical cycles [31,32]. For example, the water isotope-based damming effect of intensive reservoirs have not only influenced the water cycle [33], but also changed the bio-community structure within the watershed. A long-term (>30 years) observation found that the cascade reservoir significantly increased the amount of bioavailable nutrients (mainly N and P) due to the high density of phytoplankton induced by HRT and the nitrate reduction caused by deep hypoxia (nitrate to ammonium) [34]. As another important nutrient, carbon transport and biogeochemistry are also highly controlled by reservoir-influenced hydrodynamic conditions (mainly HRT) [35]. Moreover, the studies on HMs in dissolved loads and sediments exhibited obvious spatial variations of HM concentrations in typical commissioned reservoirs from upstream to downstream [36,37]. Since 2003, the operation of impoundment of the TGR has notably changed the hydrodynamic conditions, which directly led to the alteration of HM distribution and circulation. Therefore, continuous reporting of the current status, contamination, risks, and sources of HMs in the TGR region and the reasonable analysis are necessary.

To determine the distribution, status, and sources of HMs of river water in the upper reaches of the Three Gorges Reservoir in the Yangtze River watershed, 57 river water samples were systematically collected from June to July 2020 in the first flood season after the pandemic. The aims of this study were: (1) to clarify the distribution, status, and contamination of HMs in the upper TGR area; (2) to distinguish the potential sources of

HMs; (3) to evaluate water quality and HM-related risks. This study delivers basic data for better management in the upper TGR region and provides a reference for hydropower development and related eco-environmental issues in riverine systems.

2. Materials and Methods

2.1. Study Region

The TGR, the largest hydropower engineering project in the world, was constructed on the Yangtze River (within Yichang City, Hubei Province) with a dam of 2.3 km length and 185 m height [38]. The upper TGR region, from Chongqing City to Yichang City (660 km) was covered by reservoir water (Figure 1) [39]. There are More than 20 cities and counties located within the upper TGR region [21], which are potential pollution sources of HMs and threats due to urbanization and industrialization. The upper TGR region is affected by a humid subtropical monsoonal climate, with the average annual air temperature of 18.9 °C and the annual rainfall from 1000 to 1800 mm [40]. The lithology of the upper TGR region is composed of carbonate rocks, clastic rocks, metamorphic rocks, and evaporate rocks (Figure 1) [40]. The land use of the upper TGR region mainly consists of water area, urban area, grassland, unused land, cropland, and forest land.

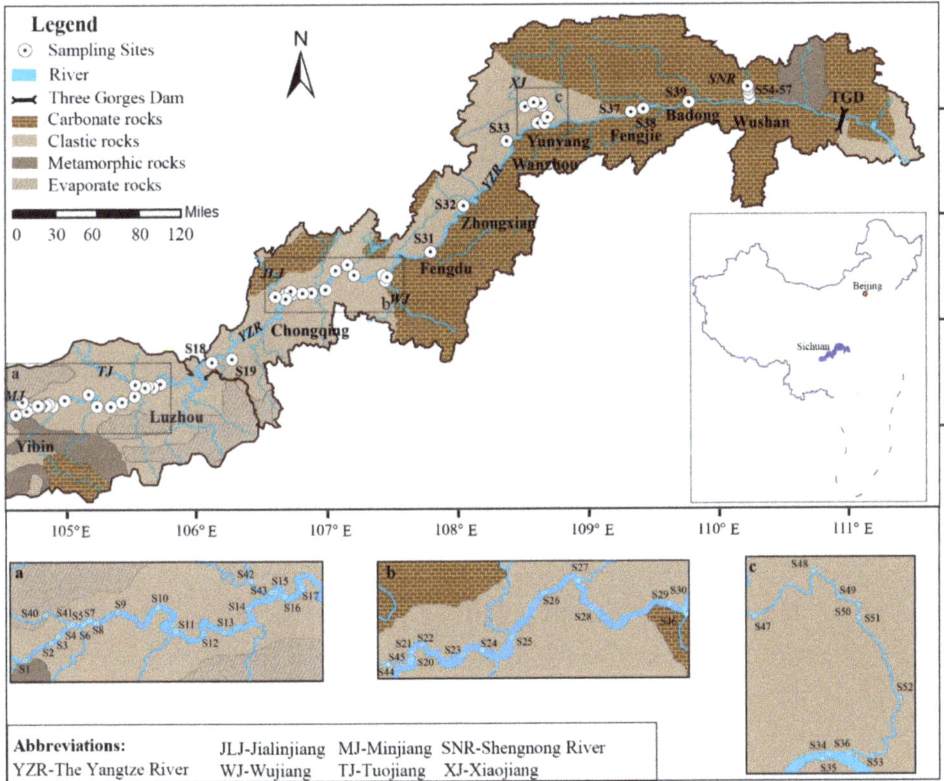

Figure 1. The lithology distribution and sampling sites of the upper TGR region.

2.2. Sampling and Chemical Analysis

A systematic sampling survey was adopted from June 25 to 21 July 2020. In the survey, 57 sampling sites including the main channel (S1–S39 and S57) and tributaries (S40–S56) were selected in the upper TGR region based on natural features (e.g., lithology) and human disturbances. In total, 57 river water samples were collected at the depth of ~50 cm. Then, 0.22 µm cellulose acetate membranes were used to filter the water samples. For HMs analysis, the water samples were acidified to pH < 2 via the pre-purified nitric acid and then sealed in clean polyethylene bottles and kept in a refrigerator until measurement. Eight HMs of river water samples were selected and measured by ICP-MS (Elan DRC-e, PerkinElmer, Waltham, MA, USA), at the IGSNRR, Chinese Academy of Sciences. The analysis was conducted using replicates, standard reference materials, and procedural blanks in order to maintain the standard quality. The measured result of replicate sample suggested acceptable repeatability for HMs (the relative standard deviation was from 0.1% for Ni to 3.5% for V). The standard reference material (GSB04-1767-2004) was applied to ensure the quality assurance for HMs analysis, which presented the recovery percentage range of 98.3% (Cd) to 101.3% (V). The measurement results (all HMs) of procedural blanks were below the limit of detection, which also clearly proved the reliability of measurements during laboratory analysis.

2.3. Assessment Method

The Water quality index (WQI) is a useful method to assess the total quality of surface water and groundwater, particularly for drinking water [25,41]. The WQI is the sum of the water quality index of individual variable (WQI_i) and can be calculated as Equation (1):

$$WQI = \Sigma[W_i \times (C_i/S_i) \times 100] \tag{1}$$

where the W_i is the relative weight ($W_i = w_i/\Sigma w_i$). W_i ranges from 1 (minimum) to 5 (maximum) based on the relative important effects of variables on human beings and the related aquatic ecological effect [42]. C_i and S_i are the concentrations/limited concentrations of variables in river water and drinking water guidelines. The limit values and weights are listed in Table 1. Here, 17 variables, including 10 water quality parameters (EC, TDS, F, Cl, NO_3-N, SO_4, Na, K, Mg, and Ca) [40] and 7 heavy metals (excluding V due to no official limit value) were incorporated in the WQI calculation. The pH of river water was not applied in the WQI calculation due to the fact that all measured pH levels were within the allowable limit (6.5–8.5).

Table 1. Relative weight and limit values of individual variable, and the calculated average WQI_i.

Variable	Drinking Water Guidelines [a]	Weight (w_i)	Relative Weight (w_i)	Average WQI_i
EC	1500 µS/cm	4	0.062	1.93
TDS	1000 mg/L	4	0.062	1.62
F	1 mg/L	5	0.077	1.97
Cl	250 mg/L	3	0.046	0.57
NO_3-N	10 mg/L	5	0.077	1.40
SO_4	250 mg/L	5	0.077	0.59
Na	200 mg/L	3	0.046	0.47
K	12 mg/L [b]	2	0.031	0.76
Mg	50 mg/L [b]	2	0.031	0.63
Ca	75 mg/L [b]	2	0.031	1.72
Ni	20 µg/L	5	0.077	0.54
Cu	1000 µg/L	2	0.031	0.00
Zn	1000 µg/L	3	0.046	0.00
As	10 µg/L	5	0.077	1.13
Mo	70 µg/L	5	0.077	0.10
Cd	5 µg/L	5	0.077	0.01
Pb	10 µg/L	5	0.077	0.02

Notes: [a] According to the Chinese drinking water standards (GB 5749-2006); [b] from [42].

To evaluate the health risk of HMs in the upper TGR area, the widely used hazard quotient (HQ) and hazard index (HI) were calculated as in previous studies [41–43]. Two main exposure pathways, namely ingestion and dermal absorption [44], were integrated in the HQ and HI calculations. HQ is defined as the ratio between exposure dose (single pathway) and reference dose (RfD), while HI s the sum of all pathways' HQs. If the value HQ or HI is >1, the human health risk/adverse effects are non-negligible. The calculation of HQ and HI were as follows [25] Equations (2)–(6):

$$ADD_{ingestion} = (C_w \times IR \times EF \times ED)/(BW \times AT) \tag{2}$$

$$ADD_{dermal} = (C_w \times SA \times K_p \times ET \times EF \times ED \times 10^{-3})/(BW \times AT) \tag{3}$$

$$HQ = ADD/RfD \tag{4}$$

$$RfD_{dermal} = RfD \times ABS_{GI} \tag{5}$$

$$HI = \Sigma HQs \tag{6}$$

$ADD_{ingestion}$ and ADD_{dermal} are the mean daily doses by two exposure pathways. C_w, BW, IR, EF, ED, AT, SA, ET, K_p, RfD, and ABS_{GI} are the river metal concentration, body weight of adults/children, ingestion rate, exposure frequency, exposure duration, average time, exposed skin area, exposure time, dermal permeability coefficient (in water) of metal, reference dose, and gastrointestinal absorption factor [25,45], respectively. The detailed values and units of these parameters were obtained from previous studies [25,45,46] and can be found in Table S1.

2.4. Software

The principal component and correlation analyses (PCA/CA) were carried out for statistics and potential HMs sources identification using SPSS 21.0. The detailed operation information of SPSS can be seen in [20,47]. All data were graphed using Microsoft Office 2010 and Origin 8.1.

3. Results and Discussion

3.1. HMs Concentrations and Distribution

3.1.1. HMs Concentrations

The statistical results of selected HM concentrations in the upper TGR area are displayed in Table 2. According to the Kolmogorov–Smirnov statistics test, most HM concentrations are normally distributed, suggesting that the average concentration of each HM is suitable for comparison. This is also supported by the similar mean values and median values of these HMs (Table 2). Therefore, the mean concentrations (in μg/L) of eight HMs in the upper TGR area decreased in the order: As (1.46), V (1.44), Ni (1.40), Mo (0.94), Cu (0.86), Zn (0.63), Pb (0.03), and Cd (0.01). As, V, and Ni are the three most abundant metals in the study area. All HM concentrations are in the range of the allowable concentrations of corresponding metals for drinking purposes recommend by the drinking water guidelines of China and the World Health Organization (WHO), except for V due to the lack of limit value (Table 2). Cu, Zn, As, Cd, and Pb are also of Grade I of Chinese surface water standard (Table 2), suggesting very clean water from the perspective of these metals in the upper TGR area. It is noteworthy that the maximum concentration of As (3.23 μg/L) is relatively close to the limit value of Chinese drinking water guidelines, which can be defined as a potential pollutant at the corresponding sampling site, revealing the relatively high human emissions of the local region [20], such as agriculture-derived As emission [48]. Compared to the background values of eight HMs in the source area of the YRW [49], the V and Ni concentrations in the Upper TGR area are significantly high (Table 2), i.e., 8.0 and 7.8 times higher than the background values. Moreover, the concentrations of Cu, As, and Mo are also relatively high, i.e., 1.4, 1.7, and 1.8 times the background values, revealing the potential anthropogenic inputs of these metals from the source area to the TGR. For historical comparison, we compared our results with the published data of the study area from 2007 to 2015. The Ni concentration in 2020 (1.4 µg/L) was comparable to the historical data, slightly higher than that in 2007 (1.2 µg/L) and 2012 (1.3 µg/L) [28,50]. The Pb concentration in 2020 (0.03 µg/L) significantly decreased from 2007 (0.86 µg/L) but was similar to the one in 2015 (0.04 µg/L) [28,51]. The concentrations of Cu (0.86 µg/L), Zn (0.63 µg/L), and Cd (0.01 µg/L) in 2020 also declined relative to those of Cu (1.2–10.4 µg/L), Zn (3.3–13.0 µg/L), and Cd (0.09–1.48 µg/L) during 2007–2015 [28,50–52]. The As concentration in 2020 (1.46 µg/L) was obviously decreased from 2007 to 2012 (2.1–7.2 µg/L), but on par with that of 2013 (1.5 µg/L) [28,50,52].

Table 2. HM concentrations, pH, electric conductivity, and total dissolved solids of river water in the upper TGR area and related limit values of drinking water guidelines.

	Unit	Upper TGR Area					Source Area of YRW [a]	Drinking Water Guidelines		Surface Water Standard [d] Standard [c]
		Min	Max	Mean	SD	Median		China [b] China [a]	WHO [c] WHO [b]	
V	µg/L	0.51	3.19	1.44	0.55	1.45	0.18			
Ni	µg/L	0.98	2.19	1.40	0.21	1.38	0.18	20	70	
Cu	µg/L	0.33	1.50	0.86	0.24	0.93	0.63	1000	2000	10
Zn	µg/L	0.00	7.90	0.63	1.14	0.33	0.68	1000		50
As	µg/L	0.45	3.23	1.46	0.57	1.48	0.86	10	10	50
Mo	µg/L	0.33	2.20	0.94	0.31	0.98	0.52	70	70	
Cd	µg/L	0.00	0.02	0.01	0.00	0.00	0.02	5	3	1
Pb	µg/L	0.00	0.10	0.03	0.03	0.02	0.76	10	10	10
pH [d]	—	7.26	8.19	7.81	0.17	7.81		6.5–8.5		
EC [d]	µS/cm	212	734	471	134	470				
TDS [d]	mg/L	119	410	264	75	263		1000		

Notes: [a] is Chinese drinking water standards (GB 5749-2006), [b] is WHO drinking water guidelines, [c,d] is the limited values of Grade I of Chinese surface water standard.

3.1.2. Spatial Distribution

Spatially, the concentrations of the HMseach HM along the main channel of the upper TGR area presented different degrees of variations (Figure 2). Overall, in the river water of the main channel, a gradually decreasing trendstrend of V, Cu, As, and Pb concentrations were observed from upstream to downstream (Figure 2), with higher concentrations in upstream sites (S1–S4, except for the highest Pb at S9). Moreover, the concentrations of V, Cu, and As increased again after the import of tributary XJ. In contrast, the concentrations of Ni and Zn were maintained at a stable level with relatively slight variations, and the highest concentrations of Ni and Zn were observed after the import of tributaries TJ and XJ, respectively. The Mo concentration was also maintained at a stable concentration level and subsequently declined to the lowest concentration after the import of tributary WJ, and then increased again after the import of tributary XJ. The Cd concentration was the most varied metal along the main channel with the highest concentration at S20 (JLJ import). The spatial distributions of the metals in six main tributaries (MJ, TJ, JLJ, WJ, XJ, and SNR) are shown in Figure 3. Compared to the average concentrations of HMs in tributaries, the highest V, Ni, As, and Mo were consistently observed in tributary TJ, indicating the potential input of these metal-related pollutants in TJ. The highest concentrations of Cu, Cd, and Pb were found in tributary JLJ, while the highest Zn concentration was exhibited at SNR. In contrast, the lowest concentrations of V, Cu, As, and Pb were displayed in tributary WJ, Ni in tributary MJ, and Zn, Mo, and Cd in tributary XJ. These results reveal the high heterogeneity of HM concentration distributions and the differences of HM characteristics in each tributary.

3.1.3. Potential Controlling Factors of Spatial Variations of HMs

The dissolved HMs in river water are affected by various factors. For river water pH, lower pH values could increase the competition between metals and hydrogen ions in binding sites and accelerate the dissolution of metal–carbonate complexes in suspended/bed sediments, which further release free metals into the river water [20,53]. In the upper TGR area, the lowest observed pH value was 7.26 (weakly alkaline water) without a significantly low pH value. Therefore, the pH values are not the main factor affecting the distribution of HM concentrations, which can also be supported by the insignificant correlation ($p > 0.05$) between pH values and the concentrations of individual HMs. Besides, water temperature can affect series of physicochemical processes (e.g., adsorption, formation of complexes, and ion exchange), and it resulted in the variation of dissolved heavy metal concentrations in water [23,54]. However, the significant variations generally occurred between different seasons (e.g., summer and winter). In this study, all samples were collected during the flood season (with negligible temperature variations); thus, the physicochemical process related to temperature was not the predominant factor. Moreover, the secondary tributary hydropower stations/dams in the upper TGR area have potentially changed the hydrodynamic condition, such as the hydraulic residence time (HRT) [29,32], and further caused the re-release of HMs from the suspended/bed sediments or gravels (product of soil erosion) due to the relatively long residence time [55–57]. Due to the lack of HRT-related studies, the secondary tributary hydropower stations/dams can only be considered as a potential explanation for the spatial variation of dissolved HMs, which may lead to the variation of concentration of some dissolved HMs in tributaries and even in the main channel.

In addition to the above-mentioned factors, source variation can be considered as the most crucial factor in the spatial distribution of dissolved HMs in river water, and it is a combination of the natural and anthropogenic sources [58–60]. In this study, large variations and high heterogeneity of HM concentration distributions were observed in both the main channel and tributaries (Figures 2 and 3), which could be attributed to anthropogenic inputs that alter their distributions or assuage the spatial variations via such a large catchment area with different landscape setting and geologic setting (natural source) in the upper TGR area. Natural sources of HMs are soils and rock weathering [60–62], and the anthropogenic sources mainly include the emission of industries, agriculture, and fossil fuel combustion [25,41,63], as discussed below.

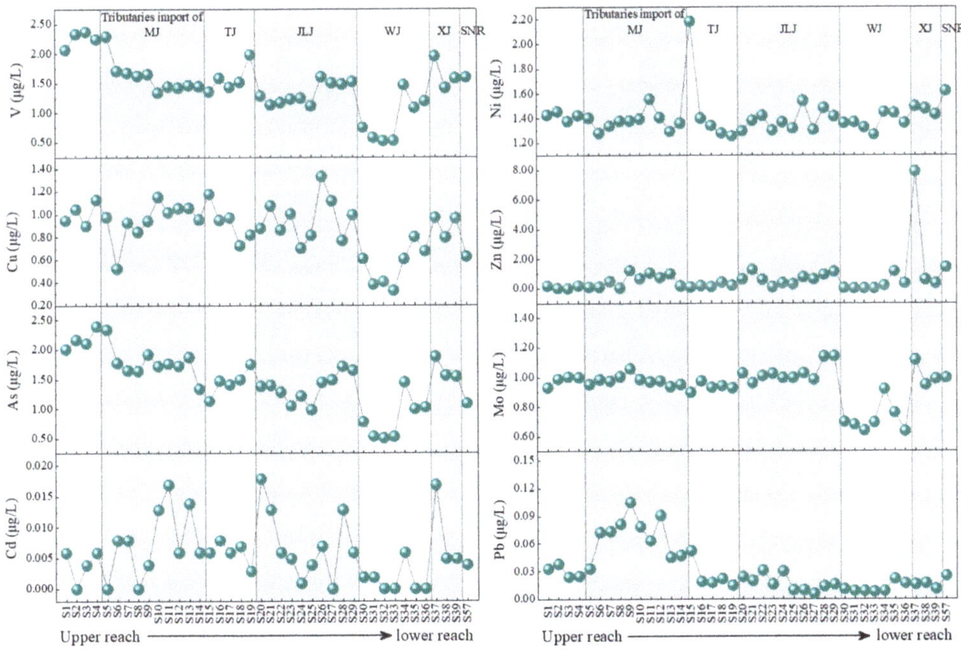

Figure 2. Spatial variations in concentrations of eight HMs at 40 sampling sites along the main channel of upper TGR area.

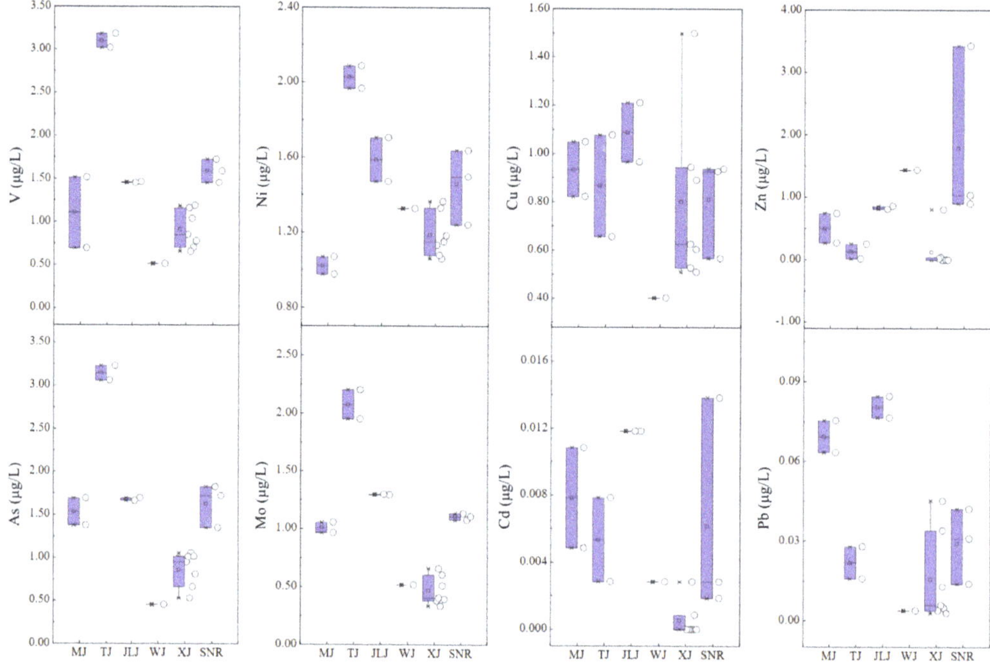

Figure 3. Spatial distribution of concentrations of eight HMs in tributaries of the upper TGR area.

3.2. Sources Identification of HMs

According to the overall coherence of the dataset, the correlation analysis could be applied to explore the relationships between different variables [64]. Here, the correlation matrix was calculated to distinguish associations among the eight HMs of river water in the upper TGR area (Figure 4a). Significant positive correlations ($p < 0.01$) were found between the V and Ni (R = 0.55), As (R = 0.0.94), and Mo (R = 0.78); Ni and Mo (R = 0.66); As and Mo (R = 0.81). Generally, the dissolved HMs with high correlation coefficients presented similar transport processes, hydrochemical behaviors, and sources [25]. Therefore, V, Ni, As, and Mo are derived from the certain co-emission sources and discharged into the river through similar chemical processes. Moreover, the moderate positive correlations were observed between Cu and V (R = 0.34), As (R = 0.44), Mo (R = 0.27), and Cd (R = 0.35); Zn and Cd (R = 0.40); Pb and As (R = 0.33), Mo (R = 0.28), and Cd (R = 0.32) (Figure 4a), indicating the interaction effect (e.g., mixing processes) of their potential co-origins and the strong spatial heterogeneity of these HMs sources. The relatively low and insignificant correlation coefficients between the pairs of some HMs (e.g., As and Zn; Figure 4a) revealed these HMs may have completely different sources.

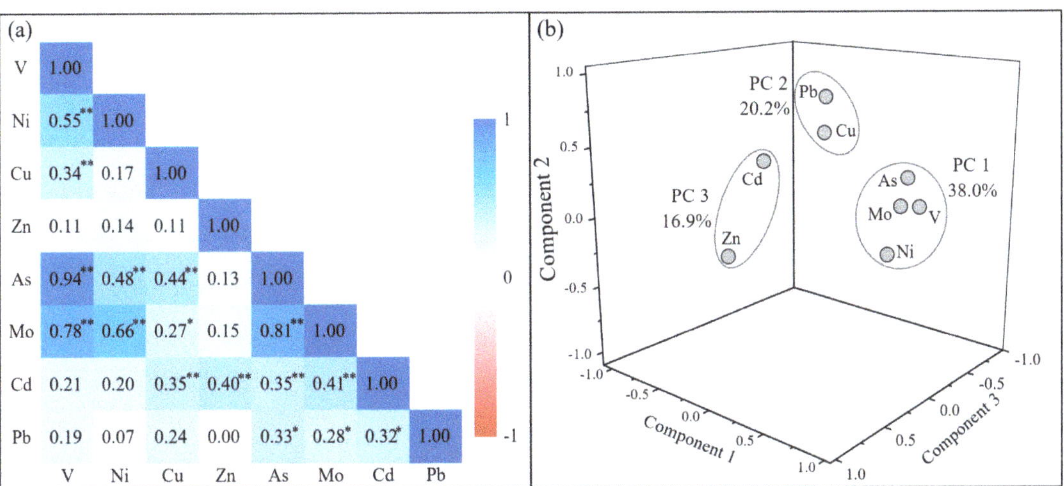

Figure 4. The correlation analysis (**a**) and principal component analysis (**b**) of eight HMs of river water in the upper TGR area. ** $p < 0.01$; * $p < 0.05$.

PCA was employed to further identify the HMs associations and their potential sources. three principal components (PC) with eigenvalues exceeding 1 were identified, which can totally account for 75.1% of the total variance. The PCA results are plotted in Figure 4b and listed in Table 3. most of the HMs showed a strong loading in their corresponding PC (loading value exceed 0.75 [65]). The first PC accounted for 38.0% of total variance with the significant loading of V (0.92), Ni (0.78), As (0.86), and Mo (0.88). The second PC explained 20.2% of total variance and was mainly contributed by Cu and Pb with high loadings of 0.63 and 0.83, respectively. The third PC including Zn (0.91) and Cd (0.67) accounted for 16.9% of total variance. However, the total PC loadings (75.1%) were slightly lower than those in previous studies, such as 86.4% and 79.3% PC loadings for 14 HMs and 13 HMs in the Dan River and Huaihe River, respectively [25,41]. The significance of KMO and Bartlett's sphericity test was less than 0.001, indicating the results of PCA were reliable. As mentioned in the correlation analysis, four HMs (V, Ni, As, and Mo) with strong positive loadings in PC1 may from the same sources; combined with the consistently high concentrations (0.94–1.46 μg/L on average, Table 2), we concluded that PC1 is mainly contributed by human inputs in the upper TGR. For example, Ni-containing pollutants were widely distributed in the wastewater of metal-processing dominated industries [63]; coal industry and agricultural activities were the main human sources of dissolved Mo in rivers [66]; As-containing pollutants were mainly derived from agricultural activities (e.g., overuse of fertilizers, pesticides, and herbicides) [48]. In contrast, Zn and Cd in PC3 were weakly correlated to HMs in PC3 (As, Mo, V, and Ni), revealing significantly different sources of HMs in PC1 and PC3. Moreover, the concentrations of Zn and Cd were relatively low (with the average concentrations lower than the background value of the source area of the YRW [49], Table 2); hence, we infer that PC3 is attributed to the contribution of natural sources (e.g., rock weathering). In addition, regarding Cu and Pb in PC2, although Cu could be derived from industrial activities and Pb could come from fossil fuel combustion [63,67], given the moderate positive correlation between these two HMs and other HMs, as well as the moderate concentration of Cu, we attribute Cu and Pb to mixed origins of human emissions and natural processes. In summary, human emission sources controlled or affected most of the HMs in the upper TGR area.

Table 3. Varimax rotated component matrix for dissolved HMs of the upper TGR area.

Variable	PC1	PC2	PC3
Eigenvalues	3.04	1.62	1.35
Variance (%)	38.0	20.2	16.9
V	0.92	0.20	−0.01
Ni	0.78	−0.10	0.16
Cu	0.23	0.63	0.16
Zn	0.08	−0.07	0.91
As	0.86	0.40	0.05
Mo	0.88	0.23	0.15
Cd	0.16	0.52	0.67
Pb	0.06	0.83	−0.06

3.3. Assessment of Water Quality and Health Risk

The WQI of different sampling sites in the study area was calculated, as plotted in Figure 5. The average WQI value of the whole upper TGR area was 13.4 ± 3.6, and the site average values of WQI in the main channel and tributaries of upper TGR area were 14.1 ± 3.4 and 11.6 ± 3.6, respectively, indicating a better water quality of tributaries. According to the result of each site, all the sampling sites had excellent water quality (WQI < 25, Figure 5) except for site S3, which presented a WQI that exceeded 25 (good water quality). For the main channel, the WQI values decreased overall from upper to lower reaches with a WQI range from 8.9 (S33) to 28.5 (S3). The WQI values of tributaries also showed a variation from 7.5 to 18.1. The lowest WQI value was observed in tributary XJ (S49), while the highest was found in tributary TJ (S42). These results reveal that the water quality of the main channel was improved with the influx of tributaries (relatively better water quality) due to the potential dilution effects and the assuaging effect of varying landscape setting [20,68]. This dilution can also be supported by the decrease of TDS from the upper (400 mg/L) to lower reaches (200 mg/L) that reported previously [40]. It is noteworthy that the WQI was mainly contributed by the water quality parameters of EC, TDS, F, NO_3-N, and Ca (with the site average WQI_i value of 1.40–1.97), while the contribution of heavy metals was relatively low, since only As showed an average WQI_i value of 1.13 (Table 1). Cl was also an important contributor of the WQI in the main channel (e.g., S1–S19, Figure 5), and the contribution of SO_4 to the WQI was also non-negligible at S10–S19, reflecting the source variation characteristics of these components (e.g., the human emission intensity and differences in rock weathering) [40,69]. Compared with large rivers of China and the polluted urban rivers, such as Zhujiang River (WQI = 1.3–43.9) [20], Langcangjiang River (WQI = 1–25) [37], and Terme River (WQI = 37.3–86.0) [42], the assessed results in the upper TGR area during the study period were pretty good. In summary, the river water in the upper TGR area met the requirements for drinking purposes based on the assessment of main water quality parameters and heavy metals, but it is still necessary to pay more attention to As.

The average concentration-based HQ and HI values of the HMs for adults and children in the main channel and tributaries in the upper TGR area were calculated, as plotted in Figure 6. The calculated $HQ_{ingestion}$, HQ_{dermal}, and HI values of all HMs for adults and children were less than 1, revealing that the investigated metals exposure via ingestion and dermal absorption were all below the harmful level. the health effects of the studied HMs on human beings are limited. Among the calculated results, the $HQ_{ingestion}$ values of most metals (except V) contributed more to HI than HQ_{dermal} values (Figure 6). It is noteworthy that the calculated HQ and HI values of all metals for children were greater than those for adults, indicating that a higher potential exposure risk of HMs for children existed, which was similar to the calculated results in other Chinese large rivers, such as Zhujiang River and Lancangjiang River [20,37]. There were no obvious differences between the HI and HQ values of the main channel and tributaries due the similar average HM concentrations. Moreover, a previous study suggested that the potentially risky effects may also occur if

HI for children is >0.1 [44]. The HI values of V and As for children exceeded 0.1 in this study (Figure 6b,d); therefore, the risk of these two metals were not completely negligible, e.g., the nephrotoxic and hepatotoxic properties and reproductive system toxicity of V [70] and the nervous system toxicity of As [45]. Thus, regular observation and appropriate measures need to be carried out to maintain the water quality in the upper TGR area and further provide better support of water resources for the development of the within-watershed economy.

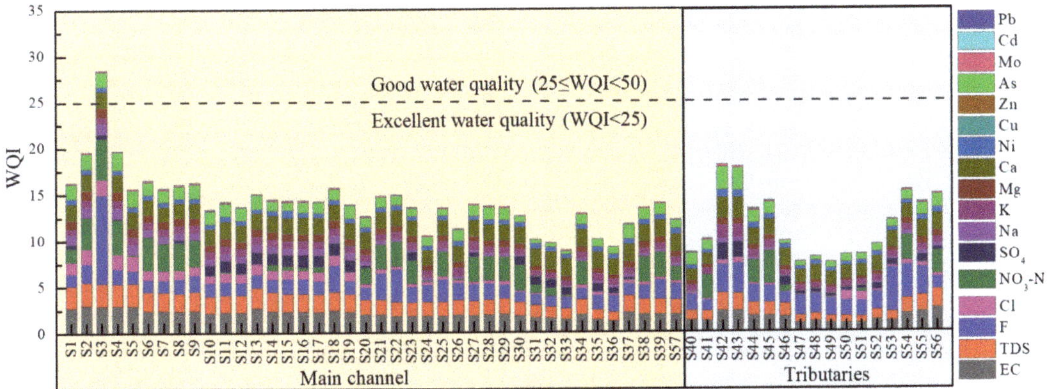

Figure 5. The spatial variations in the WQI along the main channel and tributaries of the upper TGR area.

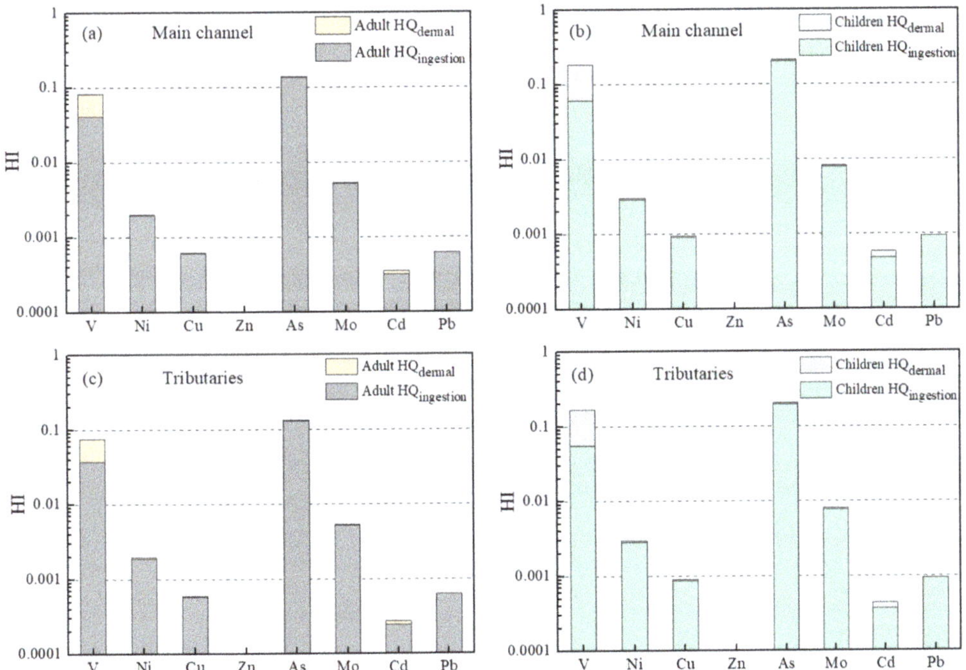

Figure 6. The HQ and HI values of HMs in the main channel and tributaries of the upper TGR area for adults and children, (**a**) HI values for adult in main channel, (**b**) HI values for children in main channel, (**c**) HI values for adult in tributaries, (**d**) HI values for children in tributaries.

4. Conclusions

The concentrations of heavy metals of river water in the upper TGR area exhibited significant spatial heterogeneity in both the main channel and tributaries. As, V, and Ni were the most abundant metals with the mean concentrations exceeding 1.40 μg/L, which were about eight times higher than the background values of the source area of the Yangtze River. All heavy metal concentrations were within the allowable values of drinking water guidelines. The water quality and health risk assessment indicated a good water quality and low risk level in the study area. Source identification suggested that V, Ni, As, and Mo were mainly contributed by anthropogenic inputs, Cu and Pb were primarily attributed to the contribution of mixed origins of human emissions and natural processes, and Zn and Cd were mainly controlled by the natural sources. Overall, good water quality was observed in the upper TGR area during the flood season, but the estimation of HM fluxes and the seasonal variations of HMs are still unclear. Moreover, the understanding of water/particle interaction of HMs (in dissolved and suspended loads) is also limited. Thus, high-frequency sampling (seasonally or monthly sampling) of both water and suspended load samples is a future concern.

Supplementary Materials: The following are available online at https://www.mdpi.com/article/10.3390/w13152078/s1, Table S1: The detailed values and units of the parameters for the calculation of hazard quotient (HQ) and hazard index (HI).

Author Contributions: Conceptualization, J.Z. and G.H.; Data curation, J.Z. and G.H.; Formal analysis, J.Z. and G.H.; Funding acquisition, G.H., M.H. and Y.W.; Investigation, M.H. and Y.W.; Methodology, J.Z. and G.H.; Project administration, G.H.; Resources, G.H.; Software, J.Z. and G.H.; Supervision, G.H.; Validation, J.Z. and G.H.; Visualization, J.Z., G.H. and D.W.; Writing-original draft, J.Z., G.H., J.L. and S.Z.; Writing-review & editing, J.Z. and G.H. All authors have read and agreed to the published version of the manuscript.

Funding: This research was funded by the National Natural Science Foundation of China (No.U1802241 and No.41325010) and the State Key Laboratory of Simulation and Regulation of Water Cycle in River Basin (SKL2020TS07).

Institutional Review Board Statement: Not applicable.

Informed Consent Statement: Not applicable.

Data Availability Statement: The data presented in this study are available on request from the corresponding author.

Acknowledgments: The authors gratefully acknowledge the group members of Mingming Hu for their assistance with field sampling.

Conflicts of Interest: The authors declare no conflict of interest.

References

1. Gao, J.; Li, C.; Zhao, P.; Zhang, H.; Mao, G.; Wang, Y. Insights into water-energy cobenefits and trade-offs in water resource management. *J. Clean. Prod.* **2019**, *213*, 1188–1203. [CrossRef]
2. Wollheim, W.M.; Bernal, S.; Burns, D.A.; Czuba, J.A.; Driscoll, C.T.; Hansen, A.T.; Hensley, R.T.; Hosen, J.D.; Inamdar, S.; Kaushal, S.S.; et al. River network saturation concept: Factors influencing the balance of biogeochemical supply and demand of river networks. *Biogeochemistry* **2018**, *141*, 503–521. [CrossRef]
3. Kamyab, H.; Din, M.F.M.; Ghoshal, S.K.; Lee, C.T.; Keyvanfar, A.; Bavafa, A.A.; Rezania, S.; Lim, J.S. Chlorella Pyrenoidosa Mediated Lipid Production Using Malaysian Agricultural Wastewater: Effects of Photon and Carbon. *Waste Biomass Valoriz.* **2016**, *7*, 779–788. [CrossRef]
4. Cai, J.; Varis, O.; Yin, H. China's water resources vulnerability: A spatio-temporal analysis during 2003–2013. *J. Clean. Prod.* **2017**, *142*, 2901–2910. [CrossRef]
5. Liu, J.; Han, G. Tracing Riverine Particulate Black Carbon Sources in Xijiang River Basin: Insight from Stable Isotopic Composition and Bayesian Mixing Model. *Water Res.* **2021**, *194*, 116932. [CrossRef] [PubMed]
6. Xu, S.; Lang, Y.; Zhong, J.; Xiao, M.; Ding, H. Coupled controls of climate, lithology and land use on dissolved trace elements in a karst river system. *J. Hydrol.* **2020**, *591*, 125328. [CrossRef]

7.   Varol, M. Environmental, ecological and health risks of trace metals in sediments of a large reservoir on the Euphrates River (Turkey). *Environ. Res.* **2020**, *187*, 109664. [CrossRef]
8.   Zhang, J.; Yang, R.; Li, Y.C.; Peng, Y.; Wen, X.; Ni, X. Distribution, accumulation, and potential risks of heavy metals in soil and tea leaves from geologically different plantations. *Ecotoxicol. Environ. Saf.* **2020**, *195*, 110475. [CrossRef]
9.   Zeng, J.; Han, G. Preliminary copper isotope study on particulate matter in Zhujiang River, southwest China: Application for source identification. *Ecotoxicol. Environ. Saf.* **2020**, *198*, 110663. [CrossRef]
10. Chen, L.; Liu, J.-R.; Hu, W.-F.; Gao, J.; Yang, J.-Y. Vanadium in soil-plant system: Source, fate, toxicity, and bioremediation. *J. Hazard. Mater.* **2021**, *405*, 124200. [CrossRef]
11. Wang, J.; Wang, L.; Wang, Y.; Tsang, D.C.W.; Yang, X.; Beiyuan, J.; Yin, M.; Xiao, T.; Jiang, Y.; Lin, W.; et al. Emerging risks of toxic metal(loid)s in soil-vegetables influenced by steel-making activities and isotopic source apportionment. *Environ. Int.* **2021**, *146*, 106207. [CrossRef]
12. Chen, L.; Liu, J.; Zhang, W.; Zhou, J.; Luo, D.; Li, Z. Uranium (U) source, speciation, uptake, toxicity and bioremediation strategies in soil-plant system: A review. *J. Hazard. Mater.* **2021**, *413*, 125319. [CrossRef] [PubMed]
13. Zeng, J.; Han, G. Tracing zinc sources with Zn isotope of fluvial suspended particulate matter in Zhujiang River, southwest China. *Ecol. Indic.* **2020**, *118*, 106723. [CrossRef]
14. Visser, A.; Kroes, J.; van Vliet, M.T.H.; Blenkinsop, S.; Fowler, H.J.; Broers, H.P. Climate change impacts on the leaching of a heavy metal contamination in a small lowland catchment. *J. Contam. Hydrol.* **2012**, *127*, 47–64. [CrossRef]
15. Viers, J.; Dupré, B.; Gaillardet, J. Chemical composition of suspended sediments in World Rivers: New insights from a new database. *Sci. Total Environ.* **2009**, *407*, 853–868. [CrossRef]
16. Islam, M.S.; Ahmed, M.K.; Raknuzzaman, M.; Habibullah-Al-Mamun, M.; Islam, M.K. Heavy metal pollution in surface water and sediment: A preliminary assessment of an urban river in a developing country. *Ecol. Indic.* **2015**, *48*, 282–291. [CrossRef]
17. Zhang, N.; Zang, S.; Sun, Q. Health risk assessment of heavy metals in the water environment of Zhalong Wetland, China. *Ecotoxicology* **2014**, *23*, 518–526. [CrossRef] [PubMed]
18. Zeng, J.; Han, G.; Wu, Q.; Tang, Y. Heavy Metals in Suspended Particulate Matter of the Zhujiang River, Southwest China: Contents, Sources, and Health Risks. *Int. J. Environ. Res. Public Health* **2019**, *16*, 1843. [CrossRef]
19. Liang, B.; Han, G.; Liu, M.; Yang, K.; Li, X.; Liu, J. Source Identification and Water-Quality Assessment of Dissolved Heavy Metals in the Jiulongjiang River, Southeast China. *J. Coast. Res.* **2020**, *36*, 403–410. [CrossRef]
20. Zeng, J.; Han, G.; Wu, Q.; Tang, Y. Geochemical characteristics of dissolved heavy metals in Zhujiang River, Southwest China: Spatial-temporal distribution, source, export flux estimation, and a water quality assessment. *PeerJ* **2019**, *7*, e6578. [CrossRef]
21. Zhao, L.; Gong, D.; Zhao, W.; Lin, L.; Yang, W.; Guo, W.; Tang, X.; Li, Q. Spatial-temporal distribution characteristics and health risk assessment of heavy metals in surface water of the Three Gorges Reservoir, China. *Sci. Total Environ.* **2020**, *704*, 134883. [CrossRef] [PubMed]
22. Jiang, G.; Xu, L.; Song, S.; Zhu, C.; Wu, Q.; Zhang, L.; Wu, L. Effects of long-term low-dose cadmium exposure on genomic DNA methylation in human embryo lung fibroblast cells. *Toxicology* **2008**, *244*, 49–55. [CrossRef]
23. Iwashita, M.; Shimamura, T. Long-term variations in dissolved trace elements in the Sagami River and its tributaries (upstream area), Japan. *Sci. Total Environ.* **2003**, *312*, 167–179. [CrossRef]
24. Thévenot, D.R.; Moilleron, R.; Lestel, L.; Gromaire, M.-C.; Rocher, V.; Cambier, P.; Bonté, P.; Colin, J.-L.; de Pontevès, C.; Meybeck, M. Critical budget of metal sources and pathways in the Seine River basin (1994–2003) for Cd, Cr, Cu, Hg, Ni, Pb and Zn. *Sci. Total Environ.* **2007**, *375*, 180–203. [CrossRef]
25. Wang, J.; Liu, G.; Liu, H.; Lam, P.K. Multivariate statistical evaluation of dissolved trace elements and a water quality assessment in the middle reaches of Huaihe River, Anhui, China. *Sci. Total Environ.* **2017**, *583*, 421–431. [CrossRef]
26. Xiao, J.; Jin, Z.; Wang, J. Geochemistry of trace elements and water quality assessment of natural water within the Tarim River Basin in the extreme arid region, NW China. *J. Geochem. Explor.* **2014**, *136*, 118–126. [CrossRef]
27. Shil, S.; Singh, U.K. Health risk assessment and spatial variations of dissolved heavy metals and metalloids in a tropical river basin system. *Ecol. Indic.* **2019**, *106*, 105455. [CrossRef]
28. Yang, Z.; Xia, X.; Wang, Y.; Ji, J.; Wang, D.; Hou, Q.; Yu, T. Dissolved and particulate partitioning of trace elements and their spatial–temporal distribution in the Changjiang River. *J. Geochem. Explor.* **2014**, *145*, 114–123. [CrossRef]
29. Yang, S.L.; Milliman, J.D.; Li, P.; Xu, K. 50,000 dams later: Erosion of the Yangtze River and its delta. *Glob. Planet. Chang.* **2011**, *75*, 14–20. [CrossRef]
30. Li, Q.; Yu, M.; Lu, G.; Cai, T.; Bai, X.; Xia, Z. Impacts of the Gezhouba and Three Gorges reservoirs on the sediment regime in the Yangtze River, China. *J. Hydrol.* **2011**, *403*, 224–233. [CrossRef]
31. Maavara, T.; Chen, Q.; Van Meter, K.; Brown, L.E.; Zhang, J.; Ni, J.; Zarfl, C. River dam impacts on biogeochemical cycling. *Nat. Rev. Earth Environ.* **2020**, *1*, 103–116. [CrossRef]
32. Wang, W.-F.; Li, S.-L.; Zhong, J.; Maberly, S.C.; Li, C.; Wang, F.-S.; Xiao, H.-Y.; Liu, C.-Q. Climatic and anthropogenic regulation of carbon transport and transformation in a karst river-reservoir system. *Sci. Total Environ.* **2020**. [CrossRef] [PubMed]
33. Yang, K.; Han, G.; Zeng, J.; Liang, B.; Qu, R.; Liu, J.; Liu, M. Spatial Variation and Controlling Factors of H and O Isotopes in Lancang River Water, Southwest China. *Int. J. Environ. Res. Public Health* **2019**, *16*, 4932. [CrossRef]
34. Chen, Q.; Shi, W.; Huisman, J.; Maberly, S.C.; Zhang, J.; Yu, J.; Chen, Y.; Tonina, D.; Yi, Q. Hydropower reservoirs on the upper Mekong River modify nutrient bioavailability downstream. *Natl. Sci. Rev.* **2020**. [CrossRef]

35. Wang, W.; Li, S.-L.; Zhong, J.; Li, C.; Yi, Y.; Chen, S.; Ren, Y. Understanding transport and transformation of dissolved inorganic carbon (DIC) in the reservoir system using δ13CDIC and water chemistry. *J. Hydrol.* **2019**, *574*, 193–201. [CrossRef]

36. Deng, L.; Liu, S.L.; Zhao, Q.H.; Yang, J.J.; Wang, C.; Liu, Q. Variation and accumulation of sediments and associated heavy metals along cascade dams in the Mekong River, China. *Environ. Eng. Manag. J.* **2017**, *16*, 2075–2087.

37. Liang, B.; Han, G.; Zeng, J.; Qu, R.; Liu, M.; Liu, J. Spatial Variation and Source of Dissolved Heavy Metals in the Lancangjiang River, Southwest China. *Int. J. Environ. Res. Public Health* **2020**, *17*, 732. [CrossRef]

38. Wang, H.; Sun, F.; Liu, W. Characteristics of streamflow in the main stream of Changjiang River and the impact of the Three Gorges Dam. *Catena* **2020**, *189*, 104498. [CrossRef]

39. Ma, Y.; Li, S. Spatial and temporal comparisons of dissolved organic matter in river systems of the Three Gorges Reservoir region using fluorescence and UV–Visible spectroscopy. *Environ. Res.* **2020**, *189*, 109925. [CrossRef]

40. Wang, D.; Han, G.; Hu, M.; Wang, Y.; Liu, J.; Zeng, J.; Li, X. Major Elements in the Upstream of Three Gorges Reservoir: An Investigation of Chemical Weathering and Water Quality during Flood Events. *Water* **2021**, *13*, 454. [CrossRef]

41. Meng, Q.; Zhang, J.; Zhang, Z.; Wu, T. Geochemistry of dissolved trace elements and heavy metals in the Dan River Drainage (China): Distribution, sources, and water quality assessment. *Environ. Sci. Pollut. Res.* **2016**, *23*, 8091–8103. [CrossRef]

42. Ustaoğlu, F.; Taş, B.; Tepe, Y.; Topaldemir, H. Comprehensive assessment of water quality and associated health risk by using physicochemical quality indices and multivariate analysis in Terme River, Turkey. *Environ. Sci. Pollut. Res.* **2021**. [CrossRef]

43. Liang, B.; Han, G.; Liu, M.; Li, X.; Song, C.; Zhang, Q.; Yang, K. Spatial and Temporal Variation of Dissolved Heavy Metals in the Mun River, Northeast Thailand. *Water* **2019**, *11*, 380. [CrossRef]

44. De Miguel, E.; Iribarren, I.; Chacón, E.; Ordoñez, A.; Charlesworth, S. Risk-based evaluation of the exposure of children to trace elements in playgrounds in Madrid (Spain). *Chemosphere* **2007**, *66*, 505–513. [CrossRef]

45. Wu, B.; Zhao, D.Y.; Jia, H.Y.; Zhang, Y.; Zhang, X.X.; Cheng, S.P. Preliminary Risk Assessment of Trace Metal Pollution in Surface Water from Yangtze River in Nanjing Section, China. *Bull. Environ. Contam. Toxicol.* **2009**, *82*, 405–409. [CrossRef]

46. Duan, X.-L.; Wang, Z.-S.; Li, Q.; Zhang, W.-J.; Huang, N.; Wang, B.-B.; Zhang, J.-L. Health Risk Assessment of Heavy Metals in Drinking Water Based on Field Measurement of Exposure Factors of Chinese People. *Environ. Sci.* **2011**, *32*, 1329–1339. (In Chinese)

47. Liu, J.; Han, G. Major ions and $\delta^{34}S_{SO4}$ in Jiulongjiang River water: Investigating the relationships between natural chemical weathering and human perturbations. *Sci. Total Environ.* **2020**, *724*, 138208. [CrossRef]

48. Xiao, R.; Guo, D.; Ali, A.; Mi, S.; Liu, T.; Ren, C.; Li, R.; Zhang, Z. Accumulation, ecological-health risks assessment, and source apportionment of heavy metals in paddy soils: A case study in Hanzhong, Shaanxi, China. *Environ. Pollut.* **2019**, *248*, 349–357. [CrossRef]

49. Zhang, L.C.; Zhou, K.H. Backgroud values of trace elements in the source area of the Yangtze river. *Sci. Total Environ.* **1992**, *125*, 391–404. [CrossRef]

50. Zhu, Y.; Yang, Y.; Liu, M.; Zhang, M.; Wang, J. Concentration, Distribution, Source, and Risk Assessment of PAHs and Heavy Metals in Surface Water from the Three Gorges Reservoir, China. *Hum. Ecol. Risk Assess. Int. J.* **2015**, *21*, 1593–1607. [CrossRef]

51. Lin, L.; Li, C.; Yang, W.; Zhao, L.; Liu, M.; Li, Q.; Crittenden, J.C. Spatial variations and periodic changes in heavy metals in surface water and sediments of the Three Gorges Reservoir, China. *Chemosphere* **2020**, *240*, 124837. [CrossRef] [PubMed]

52. Gao, Q.; Li, Y.; Cheng, Q.; Yu, M.; Hu, B.; Wang, Z.; Yu, Z. Analysis and assessment of the nutrients, biochemical indexes and heavy metals in the Three Gorges Reservoir, China, from 2008 to 2013. *Water Res.* **2016**, *92*, 262–274. [CrossRef]

53. Papafilippaki, A.K.; Kotti, M.E.; Stavroulakis, G.G. Seasonal variations in dissolved heavy metals in the keritis river, Chania, Greece. *Glob. NEST J.* **2008**, *10*, 320–325.

54. Bu, H.; Wang, W.; Song, X.; Zhang, Q. Characteristics and source identification of dissolved trace elements in the Jinshui River of the South Qinling Mts., China. *Environ. Sci. Pollut. Res.* **2015**, *22*, 14248–14257. [CrossRef]

55. Li, S.; Xu, Z.; Cheng, X.; Zhang, Q. Dissolved trace elements and heavy metals in the Danjiangkou Reservoir, China. *Environ. Geol.* **2008**, *55*, 977–983. [CrossRef]

56. Liu, M.; Han, G.; Li, X. Using stable nitrogen isotope to indicate soil nitrogen dynamics under agricultural soil erosion in the Mun River basin, Northeast Thailand. *Ecol. Indic.* **2021**, *128*, 107814. [CrossRef]

57. Han, G.; Tang, Y.; Liu, M.; Van Zwieten, L.; Yang, X.; Yu, C.; Wang, H.; Song, Z. Carbon-nitrogen isotope coupling of soil organic matter in a karst region under land use change, Southwest China. *Agric. Ecosyst. Environ.* **2020**, *301*, 107027. [CrossRef]

58. Mohiuddin, K.M.; Otomo, K.; Ogawa, Y.; Shikazono, N. Seasonal and spatial distribution of trace elements in the water and sediments of the Tsurumi River in Japan. *Environ. Monit. Assess.* **2012**, *184*, 265–279. [CrossRef]

59. Liu, J.; Li, S.-L.; Chen, J.-B.; Zhong, J.; Yue, F.-J.; Lang, Y.; Ding, H. Temporal transport of major and trace elements in the upper reaches of the Xijiang River, SW China. *Environ. Earth Sci.* **2017**, *76*, 299. [CrossRef]

60. Li, S.; Zhang, Q. Spatial characterization of dissolved trace elements and heavy metals in the upper Han River (China) using multivariate statistical techniques. *J. Hazard. Mater.* **2010**, *176*, 579–588. [CrossRef]

61. Krishna, A.K.; Satyanarayanan, M.; Govil, P.K. Assessment of heavy metal pollution in water using multivariate statistical techniques in an industrial area: A case study from Patancheru, Medak District, Andhra Pradesh, India. *J. Hazard. Mater.* **2009**, *167*, 366–373. [CrossRef]

62. Pekey, H.; Karakaş, D.; Bakoğlu, M. Source apportionment of trace metals in surface waters of a polluted stream using multivariate statistical analyses. *Mar. Pollut. Bull.* **2004**, *49*, 809–818. [CrossRef]

63. Wu, H.; Xu, C.; Wang, J.; Xiang, Y.; Ren, M.; Qie, H.; Zhang, Y.; Yao, R.; Li, L.; Lin, A. Health risk assessment based on source identification of heavy metals: A case study of Beiyun River, China. *Ecotoxicol. Environ. Saf.* **2021**, *213*, 112046. [CrossRef]
64. Chen, K.; Jiao, J.J.; Huang, J.; Huang, R. Multivariate statistical evaluation of trace elements in groundwater in a coastal area in Shenzhen, China. *Environ. Pollut.* **2007**, *147*, 771–780. [CrossRef] [PubMed]
65. Zeng, J.; Han, G.; Yang, K. Assessment and sources of heavy metals in suspended particulate matter in a tropical catchment, northeast Thailand. *J. Clean. Prod.* **2020**, *265*, 121898. [CrossRef]
66. Zeng, J.; Han, G.; Zhu, J.-M. Seasonal and Spatial Variation of Mo Isotope Compositions in Headwater Stream of Xijiang River Draining the Carbonate Terrain, Southwest China. *Water* **2019**, *11*, 1076. [CrossRef]
67. Li, Y.; Chen, H.; Teng, Y. Source apportionment and source-oriented risk assessment of heavy metals in the sediments of an urban river-lake system. *Sci. Total Environ.* **2020**, *737*, 140310. [CrossRef]
68. Chetelat, B.; Liu, C.Q.; Zhao, Z.Q.; Wang, Q.L.; Li, S.L.; Li, J.; Wang, B.L. Geochemistry of the dissolved load of the Changjiang Basin rivers: Anthropogenic impacts and chemical weathering. *Geochim. Cosmochim. Acta* **2008**, *72*, 4254–4277. [CrossRef]
69. Li, S.; Zhang, Q. Risk assessment and seasonal variations of dissolved trace elements and heavy metals in the Upper Han River, China. *J. Hazard. Mater.* **2010**, *181*, 1051–1058. [CrossRef]
70. Wilk, A.; Szypulska-Koziarska, D.; Wiszniewska, B. The toxicity of vanadium on gastrointestinal, urinary and reproductive system, and its influence on fertility and fetuses malformations. *Postepy Hig. I Med. Dosw.* **2017**, *71*, 850–859. [CrossRef]

water

MDPI

Article

Evidence of Anthropogenic Gadolinium in Triangle Area Waters, North Carolina, USA

Jordan M. Zabrecky [1], Xiao-Ming Liu [1,*], Qixin Wu [2] and Cheng Cao [1]

[1] Department of Geological Sciences, University of North Carolina, Chapel Hill, NC 27599, USA; jordanmz@alumni.unc.edu (J.M.Z.); ccapple@ad.unc.edu (C.C.)

[2] Key Laboratory of Karst Georesources and Environment, Ministry of Education, College of Resources and Environmental Engineering, Guizhou University, Guiyang 550025, China; qixinwu03@gmail.com

* Correspondence: xiaomliu@unc.edu

Abstract: Gadolinium (Gd), a member of the rare earth elements (REE), is becoming an increasingly observed microcontaminant in waters of developed regions. Anthropogenic Gd anomalies were first noted in 1996 and were determined to be sourced from Gd-based contrast agents used in magnetic resonance imaging (MRI). This study investigates Gd anomalies in North Carolina's Triangle Area, focusing on surrounding wastewater treatment plants (WWTPs). Samples were obtained from upstream and downstream of selected WWTPs as well as a freshwater reservoir that supplies part of the region's drinking water. The PAAS-normalized samples indicate Gd anomalies in the influent, effluent, and downstream samples. We quantify the anthropogenic Gd in wastewater samples to constitute between 98.1% to 99.8%. Sample comparisons show an average increase of 45.3% estimated anthropogenic Gd between samples upstream and downstream of WWTPs. This research contributes to the existing database demonstrating the presence of anthropogenic Gd in developed regions. Although current Gd concentrations are not near toxic levels, they should be continuously monitored as a micropollutant and serve as a wastewater tracer.

Keywords: gadolinium; rare earth elements; micropollutants; wastewater treatment; anthropogenic contaminants

Citation: Zabrecky, J.M.; Liu, X.-M.; Wu, Q.; Cao, C. Evidence of Anthropogenic Gadolinium in Triangle Area Waters, North Carolina, USA. *Water* **2021**, *13*, 1895. https://doi.org/10.3390/w13141895

Academic Editor: Maurizio Barbieri

Received: 27 May 2021
Accepted: 2 July 2021
Published: 8 July 2021

Publisher's Note: MDPI stays neutral with regard to jurisdictional claims in published maps and institutional affiliations.

1. Introduction

The rare earth elements (REEs) refer to the lanthanides group including elements Lanthanum (La) to Lutetium (Lu) as well as Scandium (Sc) and Yttrium (Y). These elements are grouped together as they tend to behave similarly in natural systems. However, the absolute concentration of individual REEs can vary by orders of magnitude in the same environment. Usually, the measured concentrations in a sample are normalized to a reference material representing its REE source, such as PAAS (Post-Archean Australian Shale) [1–3]. Undistributed REEs should produce a coherent and smooth pattern while any deviation of an element from the group can reflect processes causing the redistribution. Such deviations are defined as anomalies and can be calculated by comparing the elemental concentration to its neighbors. The development of modern technologies has led to a quickly raised demand of these previously "exotic" elements. As a result, REEs of anthropogenic origin have been increasingly found in the environment.

In 1996, the first anthropogenic REE anomaly was observed for the element Gadolinium (Gd) in rivers of developed regions in Germany [4]. Since then, numerous studies have reported similar anomalies in rivers [5–13], lakes [14], coastal waters [5,15–18], groundwater [6,9], wastewater [9,15,19], tap water [20,21], and even tap water-based beverages in fast food franchises [22] in developed regions across the world. The original study suggested magnetic resonance imaging (MRI) contrast agents as the most likely source of anthropogenic Gd in the natural environment. Gd-based compounds are the most commonly used MRI contrast agents. These water-soluble compounds are injected into

---

the body prior to select MRI exams to enhance produced scans and then urinated out of the body within the next 7 days (>90% in 12 hours [23]). As most traditional wastewater treatment processes do not remove anthropogenic Gd, it is released into natural waters.

In 2019, the Organization for Economic Co-operation and Development (OECD) reported that the United States had the second highest amount of hospital MRI and fourth highest number of hospital MRI exams completed in 2017 when compared to other countries. Between 2007 and 2017, the amount of hospital MRI scans completed in the United States increased from 91 to 111 scans per thousand population [24].

While anthropogenic Gd anomalies have been observed and studied in many locations, few studies focus specifically on anthropogenic Gd in North America despite the high number of MRI examinations [7,8,17]. In this study, we collected surface water samples upstream and downstream of select wastewater treatment plants (WWTPs) in North Carolina's "Triangle Area" as well as influent and effluent samples from two of the WWTPs. Additionally, we sampled a lake downstream of a WWTP that provides water for human use in the region. A previous study on REEs in the Neuse River documented an anthropogenic Gd component in the region [1]. This study aims to understand how WWTP effluent impacts Gd concentrations in surface waters in the area and supplement the existing evidence and analysis of anthropogenic Gd in North America.

2. Materials and Methods

2.1. Location of Study

North Carolina's "Triangle Area" is a term commonly used to refer to the state's Research Triangle Park, the largest research park in the United States, and its surrounding area that houses much of its workforce. The name refers to the shape obtained by using the three surrounding cities as points: Raleigh, Durham, and Chapel Hill. While not a particularly densely populated region when compared to some other urban areas, it holds extensive healthcare amenities and facilities. The area belongs to two river basins: the Neuse River Basin to the west and the Cape Fear River Basin to the east. It is located in North Carolina's Piedmont region which is defined primarily by Proterozoic to late Cambrian granite, gneiss, and metamorphosed epiclastic and volcaniclastic rocks [25].

Surface water samples were collected at Morgan Creek in Chapel Hill, Northeast Creek in Durham, and the Neuse River in Raleigh in December 2017 (Figure 1). These streams were sampled both upstream and downstream of the WWTPs. Each of these WWTPs represents an approximate "corner" of the North Carolina Triangle Area. Supplementary data for the same Neuse River sampling locations were also taken from a previous study with samples from December 2015 [1]. Additional surface water samples were taken from Jordan Lake and the Cape Fear River. Jordan Lake is an artificial freshwater reservoir that is supplied water by both Morgan Creek and Northeast Creek. In addition, 44% of Jordan Lake's water is allocated for human use in the Triangle Area [26]. Wastewater influent and effluent samples were obtained from Mason Farm and Triangle WWTPs located on Morgan and Northeast Creek, respectively. We were unable to obtain influent and effluent samples from the Neuse WWTP (Neuse River Resource Recovery Facility).

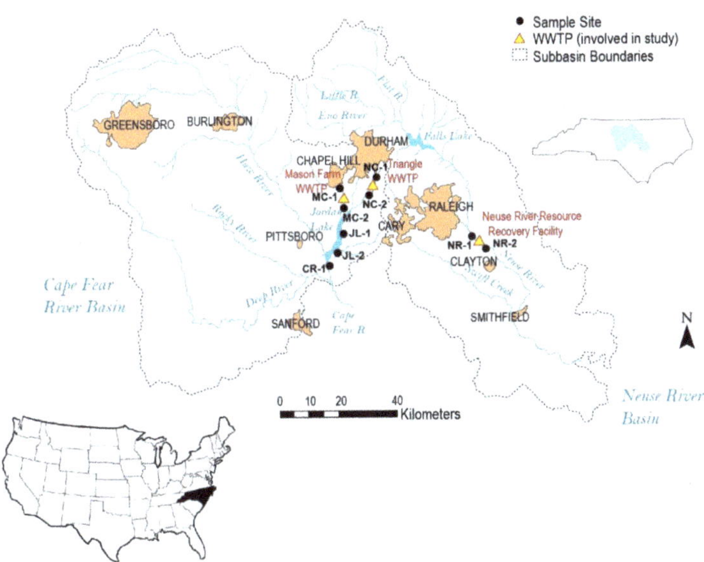

Figure 1. Map of North Carolina's Cape Fear and Neuse River Basins with the sample sites and associated WWTPs. The State of North Carolina is shaded in a base map of the United States.

2.2. Sampling and Analyses

Samples were collected in 250 ml Nalgene sampling bottles that were precleaned by 10% HCl overnight and milliQ™ water. For each sample, the bottle was rinsed three times using local river water before collecting. All samples were filtered using 0.45 μm cellulose acetate (CA) membrane filters offsite within the same day of sampling. Further, 100 mL of local river water was run through a new CA membrane filter and then discarded to prevent cross contamination from prior samplings. Between each filtering session, the filtering device was rinsed with ~750 mL of milliQ™ water. Filtered samples relocated to another clean Nalgene bottle and acidified by adding 2% concentrated nitric acid. Then, the bottles were wrapped with parafilm to minimize evaporation and then stored in a Fisher Scientific Isotemp™ refrigerator at the University of North Carolina, Chapel Hill, USA until REE analysis.

REE analysis was performed using an Agilent™ 7900 Quadrupole Inductively Coupled Plasma Mass Spectrometer (Q-ICP-MS) at the Plasma Mass Spectrometry (PMS) Laboratory at the University of North Carolina at Chapel Hill, USA. Calibration standards of 1000, 100, 10, and 5 ppt of REE concentrations were prepared from a 1000 ppm REE standard to measure REE concentrations of the dissolved loads in the sampled waters. Then, 100 ppb of Indium was used as the internal standard to correct instrumental drift. To minimize interferences, oxide presence in the plasma stream was controlled to be less than 1% in no-gas mode and 0.3% in Helium collision cell mode. To evaluate accuracy of the analysis, we analyzed a reference material for river water (SLRS-5) from the National Research Council of Canada every five samples. Repeated analysis (n = 3) of the SLRS-5 standard yielded average accuracy of <3% for La, Pr, Nd, Yb, <6% for Y, Ce, Tm, Lu, <10% for most of the rest of REE. Relative standard deviation (RSD) ranged from 1.1–15.6% with a mean RSD of 5.1%.

To characterize REE patterns, measured REE concentrations were normalized to the composition of Post-Archaean Australian Shale (PAAS) [27]. The Gd anomalies were calculated via interpolation of Sm_{SN} and Tb_{SN} [8]:

$$Gd_{SN} / Gd^{*}_{SN} = Gd_{SN} / (0.33\ Sm_{SN} + 0.67\ Tb_{SN}) \tag{1}$$

The subscript $_{SN}$ denotes sample concentration normalized to PAAS and * denotes the natural background value excluding anomalies. Anthropogenic Gd, can be calculated as follows [10]:

$$Gd^* = Gd^*_{SN} \times [Gd_{PAAS}] \tag{2}$$

$$Gd_{anth} = Gd_{measured} - Gd^* \tag{3}$$

[Gd_{PAAS}] refers to the concentration of Gd in PAAS. To account for the naturally occurring Gd anomaly, a factor of 1.1 was multiplied to Gd* in equation 3 [19].

3. Results

3.1. REE + Y Concentrations

REE concentrations obtained from all the prepared samples as well as the supplemental data from [1] are shown in Table 1. Total REE concentrations ($\sum$REE) for surface water samples range from 0.44 to 7.2 ppb with an average of 1.58 ppb. $\sum$REE concentrations from WWTP samples range from 0.27 to 0.72 ppb with an average of 0.48 ppb. Total rare earth elements and yttrium ($\sum$REY) concentrations for surface water samples range from 0.57 to 8.4 ppb with an average of 2.0 ppb. $\sum$REY concentrations for WWTP samples range from 0.40 to 0.73 ppb with an average of 0.52 ppb. The supplemental data did not include Y concentrations, so it is not included in $\sum$REY. However, Y is often included with REEs as it behaves similarly. Overall, WWTP samples have less dissolved $\sum$REE and $\sum$REY than surface water samples.

Table 1. Concentrations of rare earth elements in the sampled Triangle Area waters and WWTP influents and effluents. For WWTP samples, -I indicates the influent sample and -E indicates the effluent samples from the named WWTP.

Sample [1]	La	Ce	Pr	Nd	Sm	Eu	Gd	Tb	Dy	Ho	Er	Tm	Yb	Lu	$\sum$REE	Y	$\sum$REY
	ppt	ppt	ppt	ppt	ppt	ppt	ppt	ppt	ppt	ppt	ppt	ppt	ppt	ppt	ppb	ppt	ppb
MC-1	115	170	28.4	116	22.4	9.73	26.4	3.71	21.8	4.85	13.0	2.08	14.5	2.28	0.550	143	0.693
MC-2	184	272	44.2	181	35.6	12.5	390	5.86	42.6	9.70	30.1	4.69	33.6	5.03	1.25	272	1.52
NC-1	1376	2831	360	1412	294	72.2	289	43.0	254	44.3	113	16.6	106	14.8	7.23	1223	8.45
NC-2	346	746	91.1	355	70.8	18.2	186	10.4	61.6	11.6	28.9	4.71	30.5	13.5	1.97	387	2.36
NR-1	212	271	46.4	179	34.4	10.5	41.3	4.65	31.0	6.30	17.0	2.76	17.1	2.62	0.877	212	1.09
NR-2	276	392	61.3	241	50.5	13.3	71.5	6.56	40.2	7.99	21.4	3.17	20.2	3.28	1.21	249	1.46
NR-3	88.6	126	22.5	101	21.6	7.17	31.8	3.15	20.9	4.93	15.6	2.31	16.2	3.02	0.465	-	-
NR-4	77.6	101	20.0	90.2	19.1	6.15	61.2	2.83	18.2	4.29	13.9	2.23	15.4	2.91	0.435	-	-
JL-1	105	222	21.9	79.6	12.1	6.72	35.1	1.40	7.52	1.62	5.17	1.00	6.74	1.37	0.507	64.8	0.572
JL-2	216	492	49.3	188	39.6	10.6	122	4.17	26.6	5.03	13.5	2.13	13.2	2.46	1.19	176	1.36
CR-1	319	677	81.9	326	67.7	21.7	96.3	9.60	55.9	11.0	30.3	4.55	26.5	3.90	1.73	338	2.07
MF-I	3.16	7.20	0.647	3.51	0.704	3.59	390	0.164	0.386	0.154	0.353	0.089	2.24	0.265	0.413	4.98	0.418
MF-E	4.00	11.0	1.20	5.28	0.763	0.614	690	0.319	1.24	0.469	1.47	0.273	5.26	0.567	0.723	10.4	0.733
T-I	31.1	30.4	3.29	14.6	3.74	6.93	413	0.291	2.80	0.574	1.41	0.274	4.49	2.18	0.516	46.5	0.521
T-E	12.3	33.4	3.69	16.9	4.05	1.85	165	0.444	4.32	1.11	3.95	0.881	9.21	16.5	0.273	129	0.402

[1] MC: Morgan Creek, NC: Northeast Creek, NR: Neuse River (where 3 and 4 refer respectively to the upstream and downstream samples from [1]), JL: Jordan Lake, CR: Cape Fear River, MF: Mason Farm WWTP, T: Triangle WWTP.

Surface water samples from 2017 show enrichment of heavy REEs (HREE; Gd to Lu) in relation to lighter REEs (LREE; La to Eu) with La_{SN}/Yb_{SN} values ranging from 0.35 to 0.98 except for one sample—MF-E (Table 2). La_{SN}/Yb_{SN} is a ratio used to show relative enrichment in the lanthanides; HREE enrichment is indicated when the ratio is <1. The enrichments from the samples agree with those of the previous study of REEs in the Neuse River and is representative of REE enrichment in the geology underlying the streams of this region [1]. WWTP samples also show HREE enrichment with La_{SN}/Yb_{SN} ranging from 0.048 to 0.44. This enrichment is, on average, smaller than that of the sampled surface waters. There is one surface water sample (JL-2) that shows a slight LREE enrichment with a La_{SN}/Yb_{SN} value of 1.04.

Table 2. Select quantities for samples including La_{SN}/Yb_{SN} used to indicate fractionation between LREE and HREE, Gd anomaly ($Gd_{SN}/Gd_{SN}*$ as quantified in Equation (1)), ppt of Gd estimated to be of anthropogenic origin (Gd_{anth}), and percent of anthropogenic Gd of total Gd in sample ($\%Gd_{anth}$). For WWTP samples, -I indicates the influent sample and -E indicates the effluent samples from the named WWTP.

Sample [1]	La_{SN}/Yb_{SN}	$Gd_{SN}/Gd_{SN}*$	Gd_{anth}	$\%Gd_{anth}$
MC-1	0.498	1.29	3.90	14.7
MC-2	0.346	12.0	354	90.9
NC-1	0.819	1.17	17.1	5.92
NC-2	0.717	3.10	120	64.5
NR-1	0.785	1.50	11.1	26.9
NR-2	0.863	1.82	28.3	39.5
NR-3	0.346	1.75	11.9	37.3
NR-4	0.319	3.78	43.4	70.9
JL-1	0.048	4.00	25.4	72.5
JL-2	0.437	4.49	91.9	75.5
CR-1	0.085	1.73	35.0	36.3
MF-I	0.987	473	389	99.8
MF-E	1.04	481	689	99.8
T-I	0.759	191	411	99.4
T-E	0.089	58.0	161	98.1

[1] MC: Morgan Creek, NC: Northeast Creek, NR: Neuse River (where 3 and 4 refer respectively to the upstream and downstream samples from [1]), JL: Jordan Lake, CR: Cape Fear River, MF: Mason Farm WWTP, T: Triangle WWTP.

3.2. PAAS-Normalization and Anthropogenic Gd

PAAS-normalized REE distributions of samples show a relatively flat pattern (Figure 2). The main exception to this is the positive Gd anomaly present in WWTP samples, samples downstream of WWTPs, and the Jordan Lake samples. The Neuse River samples show the most nonlinear patterns beyond the Gd anomaly with a small negative Ce anomaly in each sample. Both negative Ce anomalies and occasional positive Eu anomalies are common in pristine waters unaffected by anthropogenic inputs [26,27]. Unlike other studies on REEs as microcontaminants, there does not seem to be any contamination from anthropogenic La or Sm [7].

The Gd anomaly (Gd_{SN} / $Gd_{SN}*$), approximated Gd concentration (Gd_{anth}), and percent of total Gd estimated to be from anthropogenic origin ($\%Gd_{anth}$) are also shown in Table 2. The highest Gd anomalies are found in the WWTP samples ranging from 58.0 to 481 with an average of 301. Surface water sample Gd anomalies range from 1.2 to 12.0 with an average of 1.4 in the samples upstream of WWTPs and 5.2 in the downstream samples. Jordan Lake samples have anomalies of 4.0 and 4.5 with the downstream of the lake having a lower anomaly of 1.73. Estimated Gd_{anth} concentrations range from 3.9 to 689 ppt. The highest Gd_{anth} concentrations were all from WWTP samples with a range of 0.39 to 0.69 ppb. The $\%Gd_{anth}$ in these samples were estimated to be 98.1% to 99.8%. The upstream samples had the smallest percentages with estimates from 5.9% to 26.9% Jordan Lake samples contained $\%Gd_{anth}$ values of 72.5% and 75.5% with a decrease to 36.3% at the downstream.

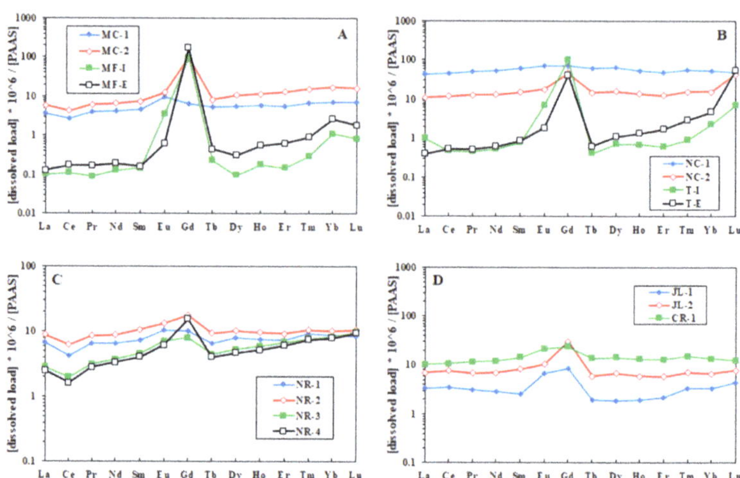

Figure 2. PAAS-normalized REE distributions for: (**A**) Morgan Creek and Mason Farm WWTP samples; (**B**) Northeast Creek and Durham WWTP sample; (**C**) Neuse River samples; and (**D**) Jordan Lake samples.

4. Discussion

4.1. Anthropogenic Gd and Its Sources

The occurrence of a small positive Gd anomaly (no higher than 1.4) is common to natural seawater due to the "tetrad effect" from Gd's half-filled 4f electron shell and REE scavenging by organic surface ligands which possess a negative Gd anomaly [28,29]. Therefore, to account for a margin of error, anomalies less than 1.4 will be considered to indicate pristine waters. In this study, only two of the Gd anomalies have a Gd anomaly of less than 1.4. These samples are the upstream samples for Northeast Creek and Morgan Creek. The Neuse River upstream shows anomaly values of 1.5 and 1.8, indicating anthropogenic Gd. This is likely because the Neuse River receives wastewater effluent from other WWTPs on streams that feed into it (such as the Smith Creek WWTP) prior to reaching the WWTP location assessed in this study. All other locations had an anomaly of at least 1.7, indicating their anomalies are all due to anthropogenic inputs. The presence of anthropogenic Gd in Jordan Lake (Gd anomalies of 4.0 and 4.5) and its downstream (Gd anomaly of 1.7) at locations roughly 14, 25, and 32 kilometers respectively downstream of the Mason Farm WWTP display the persistence of anthropogenic Gd in the environment and its efficacy as a wastewater tracer.

As surface water samples have the highest ΣREY, WWTP effluent is likely not a major contributor of overall REY concentrations in the studied streams. However, there is a large increase in estimated %Gd$_{anth}$ between upstream and downstream samples (14.7 to 90.9% for Morgan Creek; 5.9 to 64.5% for Northeast Creek, and 26.9 to 39.5% and 37.3 to 70.9% for the Neuse River). This indicates that WWTP effluent is a large contributor of anthropogenic Gd to the downstream of these rivers. Influent and effluent samples all had an estimated %Gd$_{anth}$ of at least 98.1%.

The most likely source of anthropogenic Gd in WWTP samples are Gd-based contrast agents (GBCAs). GBCAs are the most common MRI contrast agent and have been widely used since approval by the FDDA in 1988. Gd^{3+} as a free ion can compete with Ca^{2+} making it toxic to biological systems, thus it is chelated to an organic ligand to make it safe for injection [30]. Though recent studies have shown that a portion of the injected Gd can accumulate in the brain, bones, and tissues, the majority is urinated out of the body within 12 h of exposure [23,31]. Effluent released into surface waters from WWTPs contain all excreted Gd as conventional wastewater treatment processes do not remove the element. Patients are typically injected with low doses of Gd, but with the number of MRI scans conducted each year and the increase in scans per year, longitudinal studies have shown significant increases in the amount of anthropogenic Gd in waters of developed regions over the years [17,32].

4.2. Health and Ecotoxicological Effects

The first evidence of health effects from usage of GBCAs were noticed 18 years after their introduction. While GBCAs are not considered to be harmful, repeated injection can lead to noticeable Gd retention in the body and cause nephrogenic systemic fibrosis (NSF) [31,33,34]. With regards to Gd as a potential anthropogenic contaminant in natural waters, not much is known. Natural Gd in surface water is partially removed from the dissolved pool of water due to its tendency to coagulate into a colloid in low salinity waters [16]. As GBCAs injected into the body must be highly stable, anthropogenic Gd in the environment should likely be nonreactive in comparison and stay in the dissolved pool. For example, mussels in locations downstream of WWTPs have been shown to incorporate anthropogenic Sm and La into their shells, but not anthropogenic Gd, suggesting that it is not bioavailable in comparison to geogenic Gd [35]. This distinction along with GBCAs' relatively long half-life of several weeks explains how anthropogenic Gd can be used as a tracer for wastewater [36].

There are few studies that investigate the health effects of Gd alone. Most studies focus on other REEs such as Sm or La or investigate REE patterns. Some recent studies reported that REEs can accumulate in humans due to ingestion of food with high REE concentrations in contaminated soil [37,38]. Thus, drinking water with high REE concentrations could also be a potential source of REE accumulation. While anthropogenic Gd has not been bioavailable in streams, LREEs have been found to bioconcentrate in carps and specific REEs have been shown to be toxic in high enough concentrations to various organisms [39–41]. However, our sample with the highest $\sum$REE is magnitudes below one of the lowest REE concentration limits (1.39ppm of La) where ecotoxicological effects were found [40]. While the waters sampled in this study do not approach levels that cause known ecotoxicological effects, at least one other study has noted $\sum$REY levels above such limits [10]. Thus, especially with Jordan Lake as a crucial source of drinking water, Gd and other REY should be monitored as a future micropollutant in the region.

5. Conclusions

Few studies exist that show the presence of anthropogenic Gd in North American waters. This study analyzed dissolved Gd concentrations in rivers of North Carolina's Triangle Area with attention to WWTPs. Samples upstream and downstream of WWTPs were obtained as well as samples of effluent and influent of two of the three selected WWTPs. We found Gd anomalies in all samples but two of the upstream samples from pristine streams. A large change in the percentage of anthropogenic Gd (%Gd$_{anth}$) exists from upstream to downstream, demonstrating WWTP effluents large contribution of anthropogenic Gd to natural waters. The presence of anthropogenic Gd anomaly in samples taken from over 30 kilometers downstream of a WWTP indicates anthropogenic Gd's persistence in the environment and its utility as a wastewater tracer. Despite that there are no current health hazards or ecotoxicological effects at the measured concentrations in this region, Gd anomaly and total REE concentrations should be monitored because of their ability to have toxic health effects.

Author Contributions: J.M.Z. and C.C. analyzed data and wrote the manuscript with the support and supervision of X.-M.L., Q.W. collected samples and processed them. J.M.Z. wrote the manuscript with inputs from all the authors. All authors have read and agreed to the published version of the manuscript.

Funding: This research was funded by the University of North Carolina at Chapel Hill to X.-M. Liu.

Data Availability Statement: Not Applicable.

Acknowledgments: The authors thank field sampling help from WWTP staff in the Triangle Area, North Carolina. X.-M.L. and C.C. would like to acknowledge funding support from the Department of Geological Sciences, University of North Carolina at Chapel Hill. J.Z. would also like to acknowledge the support from UNC's Institute for the Environment IDEA undergraduate research program.

Conflicts of Interest: The authors declare no conflict of interest.

References

1. Smith, C.; Liu, X.M. Spatial and temporal distribution of rare earth elements in the Neuse River, North Carolina. *Chem. Geol.* **2018**, *488*, 34–43. [CrossRef]
2. Liu, X.; Hardisty, D.S.; Lyons, T.W.; Swart, P.K. Evaluating the fidelity of the cerium paleoredox tracer during variable carbonate diagenesis on the Great Bahamas Bank. *Geochim. Cosmochim. Acta* **2019**, *248*, 25–49. [CrossRef]
3. Cao, C.; Liu, X.; Bataille, C.P.; Liu, C. What do Ce anomalies in marine carbonates really mean? A perspective from leaching experiments. *Chem. Geol.* **2020**, *532*, 119413. [CrossRef]
4. Bau, M.; Dulski, P. Anthropogenic origin of positive gadolinium anomalies in river waters. *Earth Planet. Sci. Lett.* **1996**, *143*, 245–255. [CrossRef]
5. Nozaki, Y.; Lerche, D.; Alibo, D.S.; Tsutsumi, M. Dissolved indium and rare earth elements in three Japanese rivers and Tokyo Bay: Evidence for anthropogenic Gd and In. *Geochim. Cosmochim. Acta* **2000**, *64*, 3975–3982. [CrossRef]
6. Knappe, A.; Möller, P.; Dulski, P.; Pekdeger, A. Positive gadolinium anomaly in surface water and ground water of the urban area Berlin, Germany. *Chem. Erde Geochem.* **2005**, *65*, 167–189. [CrossRef]
7. Verplanck, P.L.; Taylor, H.E.; Nordstrom, D.K.; Barber, L.B. Aqueous stability of gadolinium in surface waters receiving sewage treatment plant effluent Boulder Creek, Colorado. *Environ. Sci. Technol.* **2005**, *39*, 6923–6929. [CrossRef] [PubMed]
8. Bau, M.; Knappe, A.; Dulski, P. Anthropogenic gadolinium as a micropollutant in river waters in Pennsylvania and in Lake Erie, northeastern United States. *Chem. Erde Geochem.* **2006**, *66*, 143–152. [CrossRef]
9. Rabiet, M.; Brissaud, F.; Seidel, J.L.; Pistre, S.; Elbaz-Poulichet, F. Positive gadolinium anomalies in wastewater treatment plant effluents and aquatic environment in the Hérault watershed (South France). *Chemosphere* **2009**, *75*, 1057–1064. [CrossRef]
10. Kulaksiz, S.; Bau, M. Anthropogenic dissolved and colloid/nanoparticle-bound samarium, lanthanum and gadolinium in the Rhine River and the impending destruction of the natural rare earth element distribution in rivers. *Earth Planet. Sci. Lett.* **2013**, *362*, 43–50. [CrossRef]
11. De Campos, F.F.; Enzweiler, J. Anthropogenic gadolinium anomalies and rare earth elements in the water of Atibaia River and Anhumas Creek, Southeast Brazil. *Environ. Monit. Assess.* **2016**, *188*, 281. [CrossRef] [PubMed]
12. Song, H.; Shin, W.J.; Ryu, J.S.; Shin, H.S.; Chung, H.; Lee, K.S. Anthropogenic rare earth elements and their spatial distributions in the Han River, South Korea. *Chemosphere* **2017**, *172*, 155–165. [CrossRef] [PubMed]
13. Zhang, J.; Wang, Z.; Wu, Q.; An, Y.; Jia, H.; Shen, Y. Anthropogenic rare earth elements: Gadolinium in a small catchment in Guizhou Province, Southwest China. *Int. J. Environ. Res. Public Health* **2019**, *16*, 4052. [CrossRef] [PubMed]
14. Merschel, G.; Bau, M.; Baldewein, L.; Dantas, E.L.; Walde, D.; Bühn, B. Tracing and tracking wastewater-derived substances in freshwater lakes and reservoirs: Anthropogenic gadolinium and geogenic REEs in Lake Paranoá, Brasilia. *Comptes Rendus Geosci.* **2015**, *347*, 284–293. [CrossRef]
15. Elbaz-Poulichet, F.; Seidel, J.L.; Othoniel, C. Occurrence of an anthropogenic gadolinium anomaly in river and coastal waters of Southern France. *Water Res.* **2002**, *36*, 1102–1105. [CrossRef]
16. Kulaksiz, S.; Bau, M. Contrasting behaviour of anthropogenic gadolinium and natural rare earth elements in estuaries and the gadolinium input into the North Sea. *Earth Planet. Sci. Lett.* **2007**, *260*, 361–371. [CrossRef]
17. Hatje, V.; Bruland, K.W.; Flegal, A.R. Increases in Anthropogenic Gadolinium Anomalies and Rare Earth Element Concentrations in San Francisco Bay over a 20 Year Record. *Environ. Sci. Technol.* **2016**, *50*, 4159–4168. [CrossRef] [PubMed]
18. Andrade, R.L.B.; Hatje, V.; Pedreira, R.M.A.; Böning, P.; Pahnke, K. REE fractionation and human Gd footprint along the continuum between Paraguaçu River to coastal South Atlantic waters. *Chem. Geol.* **2020**, *532*, 119303. [CrossRef]
19. Lawrence, M.G.; Ort, C.; Keller, J. Detection of anthropogenic gadolinium in treated wastewater in South East Queensland, Australia. *Water Res.* **2009**, *43*, 3534–3540. [CrossRef]
20. Kulaksiz, S.; Bau, M. Anthropogenic gadolinium as a microcontaminant in tap water used as drinking water in urban areas and megacities. *Appl. Geochem.* **2011**, *26*, 1877–1885. [CrossRef]
21. Tepe, N.; Romero, M.; Bau, M. High-technology metals as emerging contaminants: Strong increase of anthropogenic gadolinium levels in tap water of Berlin, Germany, from 2009 to 2012. *Appl. Geochem.* **2014**, *45*, 191–197. [CrossRef]
22. Schmidt, K.; Bau, M.; Merschel, G.; Tepe, N. Anthropogenic gadolinium in tap water and in tap water-based beverages from fast-food franchises in six major cities in Germany. *Sci. Total Environ.* **2019**, *687*, 1401–1408. [CrossRef]
23. Aime, S.; Caravan, P. Biodistribution of gadolinium-based contrast agents, including gadolinium deposition. *J. Magn. Reson. Imaging* **2009**, *30*, 1259–1267. [CrossRef]
24. OECD. *Health at a Glance 2019: OECD Indicators*; OECD Publishing: Paris, France, 2019.

25. Stuckey, J. *North Carolina: Its Geology and Mineral Resources*; Department of Conservation and Development: Raleigh, NC, USA, 1965.
26. NCDEQ Jordan Lake Water Supply Allocation. Available online: https://deq.nc.gov/about/divisions/water-resources/planning/basin-planning/map-page/cape-fear-river-basin-landing/jordan-lake-water-supply-allocation/jordan-lake-water-supply-allocation-background-info (accessed on April 2020).
27. McLennan, S.M. Rare earth elements in sedimentary rocks: Influence of provenance and sedimentary processes. *Rev. Mineral. Geochem.* **1989**, *21*, 169–200.
28. De Baar, H.J.W.; Brewer, P.G.; Bacon, M.P. Anomalies in rare earth distributions in seawater: Gd and Tb. *Geochim. Cosmochim. Acta* **1985**, *49*, 1961–1969. [CrossRef]
29. Lee, J.H.; Byrne, R.H. Complexation of trivalent rare earth elements (Ce, Eu, Gd, Tb, Yb) by carbonate ions. *Geochim. Cosmochim. Acta* **1993**, *57*, 295–302.
30. Sherry, D.; Caravan, P.; Lenkinski, R.E. A primer on gadolinium chemistry. *J. Magn. Reson. Imaging* **2010**, *30*, 1240–1248. [CrossRef] [PubMed]
31. Tedeschi, E.; Caranci, F.; Giordano, F.; Angelini, V.; Cocozza, S.; Brunetti, A. Gadolinium retention in the body: What we know and what we can do. *Radiol. Med.* **2017**, *122*, 589–600. [CrossRef] [PubMed]
32. Inoue, K.; Fukushi, M.; Furukawa, A.; Sahoo, S.K.; Veerasamy, N.; Ichimura, K.; Kasahara, S.; Ichihara, M.; Tsukada, M.; Torii, M.; et al. Impact on gadolinium anomaly in river waters in Tokyo related to the increased number of MRI devices in use. *Mar. Pollut. Bull.* **2020**, *154*, 111148. [CrossRef]
33. Zaichick, S.; Zaichick, V.; Karandashev, V.; Nosenko, S. Accumulation of rare earth elements in human bone within the lifespan. *Metallomics* **2011**, *3*, 186–194. [CrossRef] [PubMed]
34. Sanyal, S.; Marckmann, P.; Scherer, S.; Abraham, J. Multiorgan gadolinium (Gd) deposition and fibrosis in a patient with nephrogenic systemic fibrosis-an autopsy-based review. *Nephrol. Dial. Transplant.* **2011**, *26*, 3616–3626. [CrossRef] [PubMed]
35. Merschel, G.; Bau, M. Rare earth elements in the aragonitic shell of freshwater mussel Corbicula fluminea and the bioavailability of anthropogenic lanthanum, samarium and gadolinium in river water. *Sci. Total Environ.* **2015**, *533*, 91–101. [CrossRef]
36. Holzbecher, E.; Knappe, A.; Pekdeger, A. Identification of degradation characteristics-exemplified by Gd-DTPA in a large experimental column. *Environ. Model. Assess.* **2005**, *10*, 1–8. [CrossRef]
37. Wei, B.; Li, Y.; Li, H.; Yu, J.; Ye, B.; Liang, T. Rare earth elements in human hair from a mining area of China. *Ecotoxicol. Environ. Saf.* **2013**, *96*, 118–123. [CrossRef] [PubMed]
38. Charalampides, G.; Vatalis, K.; Karayannis, V.; Baklavaridis, A. Environmental Defects and Economic Impact on Global Market of Rare Earth Metals. *IOP Conf. Ser. Mater. Sci. Eng.* **2016**, *161*, 012069. [CrossRef]
39. Hao, S.; Xiaorong, W.; Zhaozhe, H.; Chonghua, W.; Liansheng, W. Bioconcentration and elimination of five light rare earth elements in carp (*Cyprinus carpio* L.). *Chemosphere* **1996**, *33*, 1475–1483. [CrossRef]
40. Zhang, H.; He, X.; Bai, W.; Guo, X.; Zhang, Z.; Chai, Z.; Zhao, Y. Ecotoxicological assessment of lanthanum with Caenorhabditis elegans in liquid medium. *Metallomics* **2010**, *2*, 806–810. [CrossRef]
41. Cui, J.; Zhang, Z.; Bai, W.; Zhang, L.; He, X.; Ma, Y.; Liu, Y.; Chai, Z. Effects of rare earth elements La and Yb on the morphological and functional development of zebrafish embryos. *Environ. Health Toxicol.* **2012**, *24*, 209–213. [CrossRef]

Article

Modeling Management and Climate Change Impacts on Water Pollution by Heavy Metals in the Nizhnekamskoe Reservoir Watershed

Yury Motovilov * and Tatiana Fashchevskaya

Laboratory of Regional Hydrology, Water Problems Institute of Russian Academy of Sciences, 119333 Moscow, Russia; tf.ugatu@yandex.ru
* Correspondence: motol49@yandex.ru

Abstract: The semi-distributed physically based ECOMAG-HM model was applied to simulate the cycling of the heavy metals (HM) Cu, Zn, and Mn, and to identify spatial and temporal patterns of heavy metal pollution in water bodies of a large river catchment of the Nizhnekamskoe reservoir (NKR) in Russia. The main river of the catchment is the Belaya River, one of the most polluted rivers in the Southern Urals. The model was tested against long-term data on hydrological and hydrochemical monitoring of water bodies. It is shown that the pollution of rivers is formed mainly due to diffuse wash-off of metals into rivers from the soil-ground layer. Numerical experiments to assess the impact of water economic activities on river pollution were carried out by modeling scenarios of changes in the amount of metal discharged with wastewater, a disaster with a salvo discharge of pollutants, and the exclusion of anthropogenic impact on the catchment to assess self-purification of the basin. Modeling of chemical runoff in accordance with the delta-change climatic scenario showed that significant changes in water quality characteristics should not be expected in the near future up to 2050.

Keywords: ECOMAG-HM model; river basin; heavy metals; pollution modeling; climate and management change impacts

Citation: Motovilov, Y.; Fashchevskaya, T. Modeling Management and Climate Change Impacts on Water Pollution by Heavy Metals in the Nizhnekamskoe Reservoir Watershed. *Water* **2021**, *13*, 3214. https://doi.org/10.3390/w13223214

Academic Editors: Guilin Han and Zhifang Xu

Received: 27 September 2021
Accepted: 8 November 2021
Published: 12 November 2021

Publisher's Note: MDPI stays neutral with regard to jurisdictional claims in published maps and institutional affiliations.

1. Introduction

Heavy metals (HMs) are one of the most dangerous pollutants of the environment. The main anthropogenic sources of HM are various enterprises of ferrous and non-ferrous metallurgy, mining enterprises, chemical enterprises, fuel installations, cement plants, electroplating industries, and transport. HM released into the environment are not decomposed, but rather only redistributed between its individual components. HMs accumulate relatively easily in soils, but their removal is slow. For example, the period of semi-removal of zinc from the soil is up to 500 years, while copper takes up to 1500 years [1]. In micro quantities, HMs are necessary for the processes of vital activity of organisms. Excessive amounts of HM are easily accumulated by the organs and tissues of hydrobionts and humans, adversely affecting their health and increasing the environmental risks of morbidity [2,3]. One of the main factors of the temporary detoxification of natural waters from HM ions is their adsorption by suspended particles, deposition in the form of poorly soluble inorganic compounds, and burial in bottom sediments.

HMs are one of the most common pollutants in the surface waters of Russia [4]. Their content in water bodies often exceeds the maximum allowed concentrations (MAC) established by the Russian Federation (Table 1), not only for fisheries and water use, but also for drinking water. Cases of high and extremely high pollution of HMs are recorded in individual water bodies, at which their concentrations reach values of 30–50 MAC or more.

Table 1. Maximum allowed concentrations (MAC) for some heavy metals (HM) in the Russian Federation (mg·L^{-1}).

HM	MAC in Water Bodies	
	for Drinking Water	for Fisheries
Cd	0.001	0.005
Mn	0.1	0.01
Cu	1.0	0.001
As	0.01	0.05
Ni	0.02	0.01
Hg	0.0005	0.00001
Pb	0.01	0.006
Zn	1.0	0.01
Cr^{3+}	0.5	0.07
Cr^{6+}	0.05	0.02

In recent decades, a significant decrease in the amount of wastewater discharged through point sources and, accordingly, the amount of pollutants, has been observed in most river catchments of the Russian Federation [5,6]. However, the expected decrease in the HM content in many water bodies is not marked [4,7]. This discrepancy indicates the need to deepen the understanding of the influence of various processes occurring in river basins on the formation of river water quality. Such investigations are complicated by a number of circumstances. Firstly, the existing network of state hydrochemical monitoring of water bodies in Russia has large spatial and temporal sparsity, which makes it difficult to reliably identify the regularities of the hydrochemical regime and the formation of the quality of river water under the influence of changing natural and anthropogenic factors. Secondly, due to the imperfection of the accounting system of water user enterprises that provide statistical reports on wastewater discharges, it is quite difficult to correctly assess the contribution of point sources to water pollution [8–10]. Third, monitoring of diffuse sources of pollution in the catchment area is not carried out. The best indicators of diffuse pollution are medium and, especially, small rivers, but hydrochemical monitoring points are mainly located only in large rivers in places of organized wastewater discharge: Individual large industrial enterprises, thermal power plants, nuclear power plants, as well as in the locations of cities and large settlements. Prospects for studying the patterns of HM content in river basins, predicting its changes with possible changes in climate and economic activity, and planning water protection measures in low-light conditions with observational data are associated with the construction of mathematical models describing these patterns.

Modeling the water quality of river catchments has a relatively long history. Intensive development of methods for assessing river pollution associated with the use of fertilizers in agriculture began in the 1960s. These methods were based mainly on empirical relationships for calculating pollutant washouts from agricultural fields using correlations with climatic and physico-geographic parameters [11].

One of the first conceptual field-scale models of water quality was the CREAMS model (Chemicals Runoff and Erosion from Agricultural Management Systems) [12]. The model includes a description of processes in hydrology, erosion, plant, and chemistry sub-models and is used in agricultural management practices. The main limitations of the CREAMS are related to the small size of the simulated area bound by the agricultural field. The ideology of the CREAMS model was the impetus for the creation of many modifications of conceptual models, describing water quality formation in areas with nonpoint spatial pollution (reviews can be found in [13–16]).

Since the 2000s, semi-distributed models of the river runoff and water quality formation have been used to evaluate the effects of alternative management decisions on water resources and non-point source pollution in river basins. Among others, the Soil and Water Assessment Tool (SWAT) is widely used for the simulation of river discharge and pollution loads at different scales: From the continental scale [17], through the regional scale [18], to the scale of individual river catchments [19,20]. There is evidence in using semi-distributed models for the evaluation of measures to reduce contaminant loads and improve the ecological status of water bodies in Germany (SWIM, Soil and Water Integrated Model [21]; MONERIS, Modeling Nutrient Emissions in River Systems [22]; METALPOL model [23]) and in Sweden (HYPE, Hydrological Predictions for the Environment [24]) in the framework of the implementation of the European Water Framework Directive. In semi-distributed models, the watershed is divided into several subbasins—Hydrological Response Units (HRUs). Each HRU is divided into several layers, and the vertical and horizontal fluxes of water, sediment, and nutrients loadings that move into the river network are modeled. Previous authors [24] drew comparisons between several semi-distributed water quality models based on their complexity, input data constraints, and performance of the models. Overviews of the water quality formation models can be found in [25–28].

Most of the semi-distributed models of water quality formation in river basins were used mainly for biogenic elements (nitrogen and phosphorus). Much less research is devoted to modeling the formation, transport, and transformation of heavy metals in rivers, which may create serious environmental problems in the region. Such models are usually developed for mining areas [29] with widespread waste rock dumps, tailing dumps [19], and contaminated soil [30], as well as in areas with a wide spread of "dirty" industries [31]. For example, in [19], the load of HM transported by the highly contaminated mine drainage river has been assessed using a combination of the SWAT model (for hydrological simulation) and an empirical equation statistically relating local discharge with the concentration of elements. A statistical approach using a multilayer perceptron neural network model was also used in [29] to forecast the content of HM pollutants in stormwater sediments on the basis of atmospheric data and catchment physico-geographic characteristics. In [31], a detailed heavy metal transport and transformation module was developed and combined with the SWAT model for the purpose of simulating the fate and transport of metals at the watershed scale. This modification of the SWAT-HM model was calibrated and validated to simulate Zn and Cd dynamics in the watershed, which has been impacted by mining activities for decades. The authors noted that the developed model has a high potential for application in environmental risk analysis and pollution control to provide scientific support for pollution control and remediation decisions.

One physically based model of runoff and water quality formation in river basins developed in Russia is the ECOMAG-HM model (Ecological Model for Applied Geophysics–Heavy Metals) [30,32]. The model is well adapted to the description of processes in river basins with a snow-type water regime, as well as the structure and composition of hydrometeorological information in Russia. The hydrological module of the ECOMAG model has been repeatedly tested on many large river basins and is used in the practice of hydrological calculations, forecasts, and water resource management in the Russian Federation [33]. The hydrochemical module was applied to assess possible changes in river water pollution under various scenarios of the Pechenganikel mining complex in the north-west of the Kola Peninsula, which creates environmental problems in the region [30]. Besides, the model was verified using hydrochemical monitoring data in a large river basin to study the regime of HM concentrations in the watercourses of the Nizhnekamskoe reservoir (NKR) catchment in Russia (for copper [32] and zinc [34]) [35].

The main purpose of the study is to assess the ECOMAG-HM model's potential to study spatio-temporal patterns of HM runoff formation in a large river basin (with the example of the NKR basin) to quantify scenarios of the water protection measures under possible changes in water economic activity and climate. The specific objectives of the study were (a) to evaluate the capabilities of the model for assessing HM pollution of river waters in various sections of the river network to solve the problem of designing a hydrochemical monitoring network; (b) to assess the contributions of diffuse and point anthropogenic sources to HM pollution in the NKR watershed; (c) to assess changes in water quality of water bodies under various scenarios of anthropogenic load on the river basin; and (d) to assess changes in the hydrological and hydrochemical regime of rivers under a scenario of possible climatic changes. In this paper, in the section of the testing results of the model, more attention is paid to the study of the spatio-temporal patterns of manganese in river waters. The results of the copper and zinc simulations in the river system are used for comparison with the manganese regime, as well as to illustrate the HM regimes under various scenarios of changes in water economic activity and climate.

2. Materials and Methods

2.1. Study Area: The Nizhnekamskoe Reservoir Watershed

The Nizhnekamskoe reservoir was created in the valley of the Kama River (the largest tributary of the Volga River) in 1979. The area of the NKR watershed between the Nizhnekamsk (Naberezhnye Chelny city) and Votkinsk (Tchaikovsky city) hydropower plants is 186,000 km^2, most of it (142,000 km^2) occupied by the basin of the Belaya River River (Figure 1). About 2/3 of the catchment area in the western and central parts is flat territories, and the eastern part is the Ural folded mountain region. The average height of the basin is 392 m, and the highest is 1654 m (Yaman-Tau Mountain). Forest occupies about 50% of the territory.

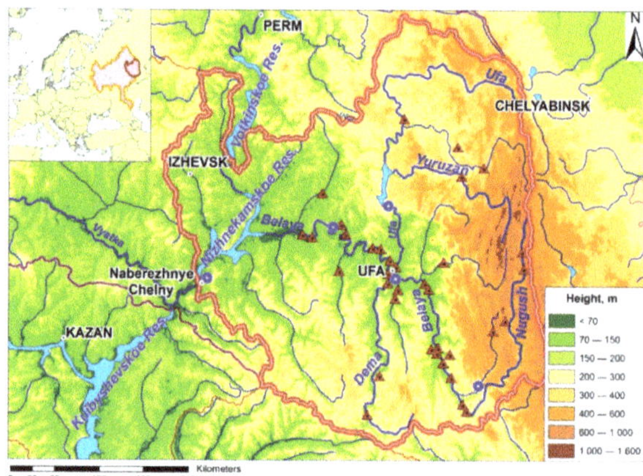

Figure 1. Map of the Nizhnekamskoe Reservoir (NKR) watershed and location of monitoring posts: Hydrological gauges (circles) and hydrochemical monitoring points (triangles). The inset shows the location of the NKR watershed (pink) in the Volga basin (gray) and Eurasian continent.

The climate of the territory is continental. There are a number of transitions from the climate of semi-arid steppe regions, where the annual precipitation ranges from 300–400 mm and the average annual air temperature is about 3 °C, to more humid areas (north-eastern and eastern mountain-forest), where the annual precipitation exceeds 600 mm and the average annual air temperature is below 1 °C. The rivers are mainly fed by snow. More than 60% of the annual runoff passes during the spring flood. The average annual lateral inflow of water to the NKR is 36.5 km^3, of which 26.1 is accounted for by the flow of the Belaya River. To compensate for the shortage of water resources, which often occurs in low-water years, about 400 reservoirs and ponds with a volume of more than 100,000 m^3, as well as many smaller ponds, operate in the basin. The largest reservoirs are Pavlovskoye on the Ufa River, Nugushskoye on the Nugush River, and Yumaguzinskoye on the Belaya River.

A feature of the geological structure of the territory is the wide distribution of ore deposits and the significant concentration of ore elements in rocks. More than 3000 deposits and manifestations of various types of minerals have been discovered in the NKR basin [36]. The soils of the catchment area (chernozems, sod-podzolic, gray forest) are characterized by a high content of humus and heavy mechanical composition. Well-drained mountain soils are common in the east of the region. Soils inherit the chemical composition of soil-forming rocks. In the soil, HMs are included in the composition of humic substances, forming strong complexes with humic and fulvic acids, and are not removed from it for hundreds and thousands of years [1]. The heavy mechanical composition of soils contributes to the increased accumulation of HM. About 60% of the studied territory is erosively dangerous land as a result of water and wind erosion. Eroded soils enriched with trace elements contribute to the entry of HMs into water bodies with sediments [36,37].

The industrial development of the region began almost 300 years ago, was associated with the development of mineral deposits, and took place without taking into account environmental restrictions. "Ancient" anthropogenic-transformed mining landscapes, modern industrial enterprises for the extraction and processing of mineral resources, large settlements, and their infrastructure objects are sources of additional metal supply to the catchment area. In the eastern part of the catchment, such sources are mining enterprises, while in the western and central parts, they are enterprises of oil production, oil refining, chemistry and petrochemistry, metallurgy, mechanical engineering and energy, and objects of storage of production and consumption waste.

As mentioned above, in recent decades, the volume of pollutants entering water bodies from controlled point sources has been decreasing for reasons that are not fully established: Perhaps due to the decrease in water consumption [5,6] or due to the decrease in the number of monitored water users providing statistical environmental reports of industry, settlements, and communal services on the 2-TP (vodkhoz) form [8]. In [38], an assessment of the correctness of the information contained in the reporting forms of 2-TP (vodkhoz) was evaluated. According to the results of the assessment, a large number of cases were revealed when the amount of metal actually contained in wastewater significantly exceeded the statistical data, and in some cases, the excess was 1–2 orders of magnitude. The revealed differences can be explained by the shortcomings of the existing system for monitoring the composition of wastewater at enterprises. In the absence of automatic continuous monitoring of the HM content in wastewater at enterprises, the assessment of the HM discharge is carried out either on the basis of individual episodic water samples, or indirectly by the number of products produced. Such approaches inevitably introduce significant errors in determining the actual HM discharges from point sources [9,10].

2.2. ECOMAG-HM Model

The semi-distributed physically based ECOMAG-HM model [30,32] was developed to simulate the spatio-temporal dynamics of hydrological and hydrochemical cycle components in large river basins. The model operates with a daily time step and consists of two main blocks: The hydrological submodel of runoff formation and the hydrochemical submodel of pollutant migration and transformation. The first submodel describes the main processes of a land hydrological cycle: Vertical fluxes (infiltration of rain and snowmelt water into the soil, percolation of water into deeper soil horizons and groundwater zone, evapotranspiration), changes in water and energy contents in different components of the geosystem (snow cover formation and melting, soil freezing and melting, surface water, soil moisture, groundwater level), and lateral fluxes (formation of surface, subsurface, groundwater flow), which form the river runoff.

The hydrochemical submodel describes the processes of migration and transformation of conservative pollutants in catchments. It accounts for pollutant accumulation on land surfaces, dissolution by melt and rainwater, penetration of dissolved forms of pollutants into soil, and interactions with soil solution and soil solid phases. Pollutants are transported by surface, subsurface, and groundwater flows to the river network and form the lateral diffuse inflow into rivers. The submodel of the transfer and transformation of pollutants in the river system include the lateral diffuse inflow and load of pollutants from point anthropogenic sources to the rivers and routed through the river network to the outlet of the catchment, taking into account the exchange of pollutants between the river water and riverbed.

The Information-Modeling Complex (IMC) ECOMAG was developed for easy use and adaptation of the model for river basins under various projects of information support in water resources management, forecasting water and hydrochemical regimes, and research. The IMC computer technology contains the calculation module of the mathematical model, databases, and instruments for information and technological support of this module. Thematic digital maps (a digital elevation model, hydrographic network, soils, and landscapes) are applied for automated division of the river basin into elementary watersheds (model cells, analog HRU in SWAT) and a schematization of the river network using the specific GIS tool Ecomag Extension. Data on hydrometeorological, hydrochemical, and water management monitoring as well as instruments for geoinformation processing of these spatial data are involved in the IMC ECOMAG for calibration of the model parameters, verification, testing of the model, and calculations for various projects.

The concept, algorithms, equations, and results of the application of the hydrological module in operative water management and forecast practice were described in many studies [33] and applied to river basins of different scales (including the largest in the northern hemisphere: The Volga, Lena, Selenga, Amur, Mackenzie, etc.) located in different geographic zones with different runoff formation conditions and hydrological regimes of water bodies. The geography of application of the hydrochemical module of the ECOMAG-HM model for environmental problems is more moderate. A description of the hydrochemical module and results of validation of the model were published in [30,32,34,35].

2.3. Data Preparation and Model Setup

For calculations on the model, boundary conditions were set in the form of daily spatial meteorological fields, concentrations of heavy metals in atmospheric precipitation, and pressure groundwater feeding the upper groundwater zone. Daily data on 56 meteorological stations located in the NKR watershed for the 1979–2011 period were used to construct meteorological fields of air temperature, precipitation, and air humidity deficit. The concentrations of metals in atmospheric precipitation were set by constant values based on the weighted average concentrations given in [39]. Furthermore, the concentrations of metals in the pressure groundwater were set as constant values, based on the data given in [40]. In addition, as information on the point anthropogenic sources of river water pollution, mean annual data on metal discharges with wastewater in 12 large settlements in the Belaya River were set on the basis of State statistical reporting forms 2-TP (vodkhoz) for the 2002–2007 period. Maps of heavy metals content in the arable soil layer on the territory of the Republic of Bashkortostan [41], which occupies a large part of the NKR catchment area, were used to set the initial conditions for the hydrochemical module of the ECOMAG-HM model.

To calibrate the model parameters and verify the hydrological module, data on daily river discharge at 5 hydrological gauges were used. The data of observations of HM concentrations at points of the Russian State Monitoring Survey on the Belaya River and its tributaries were used to calibrate the parameters and test the hydrochemical module: For copper and zinc, at 35 points for the 2004–2007 period; for manganese, at 26 points for the 2002–2007 period. As a rule, observations of the metal content in river waters were carried out once a month.

The spatial schematization of the catchment area and the river network in the NKR basin was produced using the Ecomag Extension tool on the basis of a digital elevation model. As a result, 503 elementary watersheds were identified in the catchment; their average area is about 400 km^2.

The values of most of the model parameters were set from the IMC ECOMAG databases using available maps of land surface characteristics (relief, soil, vegetation) or were determined on the basis of literary data [42]. Some parameters of the model were corrected using the calibration procedure by comparisons with daily runoff hydrographs and observations of HM concentrations in river waters at observation stations within the framework of hydrological and hydrochemical monitoring of water bodies. Numerical experiments have shown that the most sensitive parameters of the hydrological submodel are the saturated hydraulic conductivities of soils, the evaporation index, and the degree–day coefficient of snow melting. In the hydrochemical submodel, the constants of the sorption equilibrium of HM in soils and the parameter accounting for the exchange of metals between river water and bottom sediments were refined by calibration.

2.4. Scenarios of the Water Economic Activity Changes Experiments

In order to assess the impact of water economic activity changes on river water pollution, a series of numerical experiments was evaluated, in which the following scenarios were performed:

1. an increase in the amount of metal in the wastewater of all localities with controlled wastewater discharges;
2. a salvo of a significant amount of HM into the watercourse as a result of an emergency discharge in one of the localities;
3. complete exclusion of anthropogenic impact on the catchment.

2.4.1. Scenario 1. Change in the Amount of Wastewater Entering the River Waters

Water pollution was simulated under various scenarios of changes in the amount of metal discharged in wastewater at 0.1, 10, 20, 40, 60, 80, and 100 times in relation to the existing level (represented as 2-TP (vodkhoz) forms) in all localities. In fact, such changes may be due to an increase in industrial production, or in some cases, as a result of more objective and correct presentation of information in the 2-TP (vodkhoz) reports on the amounts of HM discharged by water users in some localities [9,10,38].

2.4.2. Scenario 2. The Simulation of a Disaster

The simulation of an extreme situation that led to a significant amount of HM entering the watercourse was carried out. The reason for this situation may be accidents at liquid waste storage facilities, multiple discharges of wastewater as a result of unstable operation of enterprises, unauthorized discharges of contaminated wastewater without treatment, etc. The algorithm of events was as follows. As the input to the model, various amounts of metals were dumped at a certain point in the river network during the day. Then the concentrations of metals in the river network below the emergency point were calculated.

2.4.3. Scenario 3. Exclusion of Anthropogenic Impact on the Catchment Area

The estimation of the time scale of self-cleaning of the catchment from HM in the scenario of complete exclusion of anthropogenic impact on the NKR basin was performed. It was assumed that HMs were not dumped in the wastewater, and the entry of metals into the atmosphere with industrial emissions was also excluded (the concentration of metals in precipitation was equal to zero). The concentration of metals in the pressure groundwater decreased to the lowest level in the range of typical values for the underground water of the studied region. Under these conditions, numerical experiments were carried out to estimate the dynamics of heavy metals in river waters for 400 years ahead. A series of historical data of meteorological observations at weather stations for 33 years, from 1979 to 2011, was used as a meteorological forcing. The final results of calculations of the river basin characteristics for 31 December 2011 were recorded at the starting point of 1 January 1979, and thus the calculations for the 33-year time series were repeated many times.

2.5. Description, Evaluation, and Processing of Climate Scenario Data

To construct a scenario of the future climate for the NKR basin, the trends in annual temperature, precipitation, and runoff for several meteorological and hydrological stations located in the watershed area over the period of 1956–2007 were analyzed. It was found that at the turn of the 1980s, there were changes in climate signals between the climate parameters for two periods. The average air temperature for the 1980–2007 period increased by about 1 °C compared to the previous period of 1956–1979. The increase in annual precipitation over the 1980–2007 period compared to the previous period averaged for the whole considered meteorological stations was about 10%. The difference in the mean annual runoff was about 10% in the gauges of the Belaya River during these periods (see examples in Figure 2). Modeling the water and chemical runoff for the future period, up to 2050, was carried out using the so-called "delta change" (DC) climatic scenario, assuming changes in climate parameters to be the same as for the previous observation periods. So, the obtained increments of the annual daily air temperature of 1 °C and daily precipitation multiplied to 1.1 were introduced to the corresponding historical daily series of climate parameters for the last observation period. It should be noted that similar trends in runoff changes for the region under consideration over the past decades and in the near future are given in the review [43].

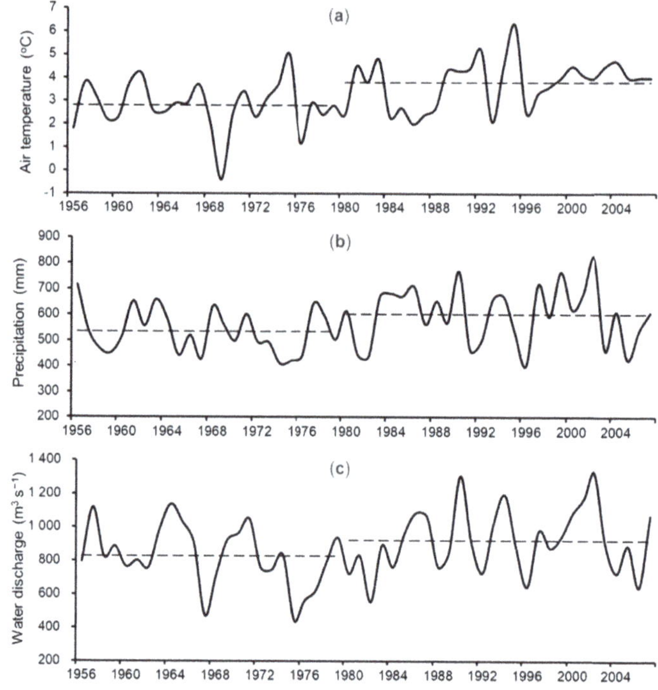

Figure 2. Changes in mean annual air temperature (**a**) and precipitation (**b**) in Ufa station and the mean annual discharges (**c**) of the Belaya River at Birsk gauge for the 1956–2007 period (dotted lines are trends).

3. Results and Discussion

3.1. ECOMAG-HM Testing of the Model

3.1.1. Model Verification

The hydrological submodel was calibrated and verified by comparing the observed and simulated runoff hydrographs at several hydrological stations in the NKR basin over the 1978–2013 period. The results of the tests showed satisfactory agreement between the observed and simulated runoff characteristics according to the Nash–Sutcliffe efficiency criterion for the daily hydrographs, and the PBIAS criterion for the annual runoff volumes [32,33]. Figure 3 shows these comparison results at four stations over the 2002–2007 period, which will be later used for the simulation of hydrochemical characteristics.

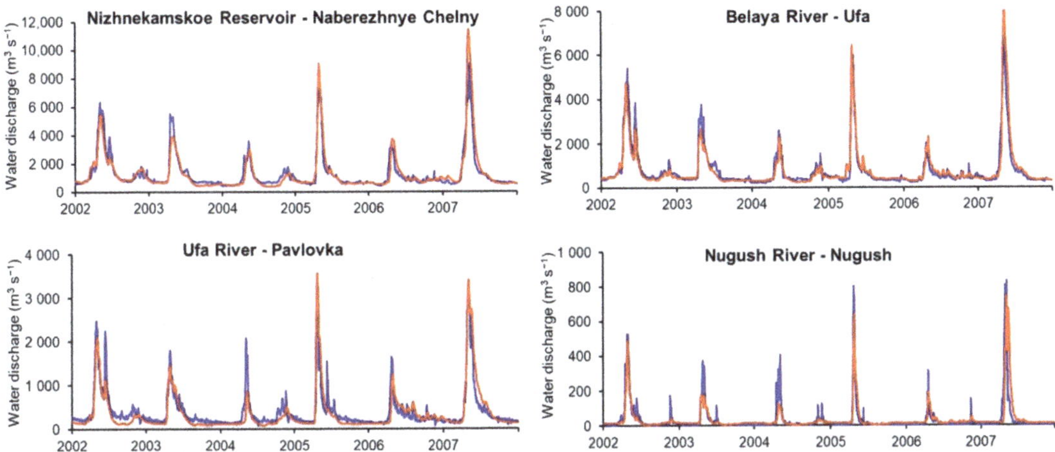

Figure 3. Observed (blue) and simulated (red) hydrographs of the Belaya River and inflow into NKR.

An additional successful test of the model for the rivers not covered by hydrometric observations was performed by comparing the map of the mean annual specific runoff estimated on the basis of averaging the simulated daily runoff in the nodes (centers of gravity) of elementary watersheds over many years with the map of specific runoff given in SNIP-2.01.14-83 [32,44]. The latter is used in Russia for the estimation of the hydrological characteristics in the design of hydraulic structures on rivers in the absence of hydrometric observations.

The hydrochemical submodel was tested by comparing observed and simulated concentrations of copper [32] and zinc [34] in different sections of rivers of the NKR basin. The rare frequency of observations at the hydrochemical monitoring points (usually up to 12 observations per year), as well as significant errors in determining HM concentrations in water samples (~50%), do not allow the full use of traditional statistical hydrological criteria for comparing the results of model calculations of intra-annual HM concentrations and episodic data of hydrochemical measurements [45]. Characteristics averaged over longer time periods can be determined more reliably based on such data.

Examples of the comparison of the intra-annual distribution of Mn concentrations (daily and averaged over quarter (three-month) periods) are shown in (Figure 4). As can be seen, seasonal changes in the Mn content are characterized by increased concentrations during high-water periods (snowmelt and rainfall runoff) and a decrease during the summer and winter low-flow periods. Correlation analysis showed that the maximal correlation coefficients, R, between the observed and simulated concentrations at all monitoring points in the basin, averaged by three months (quarters), were obtained for copper in the second quarter (the spring flood period, R = 0.55), for zinc in the second and third quarters (R = 0.50 and R = 0.69, respectively), and for manganese in all quarters (R varies in the range from 0.68 (fourth quarter) to 0.85 in the first quarter).

Spatial differences in the long-term mean annual HM concentrations in the Roshydromet hydrochemical monitoring points located in the order from the upper reaches of the Belaya River toward its inflow into the NKR differ significantly (Figure 5). Table 2 shows the correspondences between the observed and simulated HM concentrations and the range of changes in monitoring points. The correlation coefficients between simulated and measured concentrations are 0.58 for copper, 0.60 for zinc, and 0.80 for manganese.

Additional verification of the model was carried out by comparing the calculated and actual inflow of pollutants into the NKR. The model allows for calculating the balance of chemical components entering the river network and estimating the contribution of point (controlled wastewater discharges) and diffuse sources to river water pollution. As the information on point sources, yearly data on metal discharges with wastewater in 12 large localities were used. The actual inflow of metals from the catchment area to the NKR was estimated by multiplying the observed annual lateral inflow of water into the reservoir by the measured average annual concentration of metal at the last hydrochemical monitoring station on the Belaya River before its confluence with the NKR (Dyurtuli station) (Table 3).

Table 2. Observed and simulated HM concentrations and its ranges of changes in monitoring points ($\mu g \cdot L^{-1}$).

HM	Long-Term Mean Annual Concentration Averaged over Monitoring Points		Range of Changes in Long-Term Annual HM Concentrations at the Monitoring Points			
			min		*max*	
	Observ.	Simul.	Observ.	Simul.	Observ.	Simul.
Cu	3.94	4.28	2.38	2.59	6.29	7.10
Zn	4.48	4.65	2.36	2.72	9.03	10.24
Mn	84.4	77.1	42.4	42.4	188.0	116.1

Table 3. Simulated components of HM load into the river network from the NKR watershed and the actual and simulated HM load into the NKR ($t \cdot year^{-1}$).

HM	HM Load		Actual Load of HM by Wastewater	HM Setting onto Riverbed	Total HM Load into NKR	
	With Surface Water Flow	With Subsurface Water Flow			Simulated	Actual
Cu	40.3	174.3	2.3	99	118	119
Zn	35.8	197.5	8.9	144.3	98	114.5
Mn	685	3093	3.2	1134	2651	2651

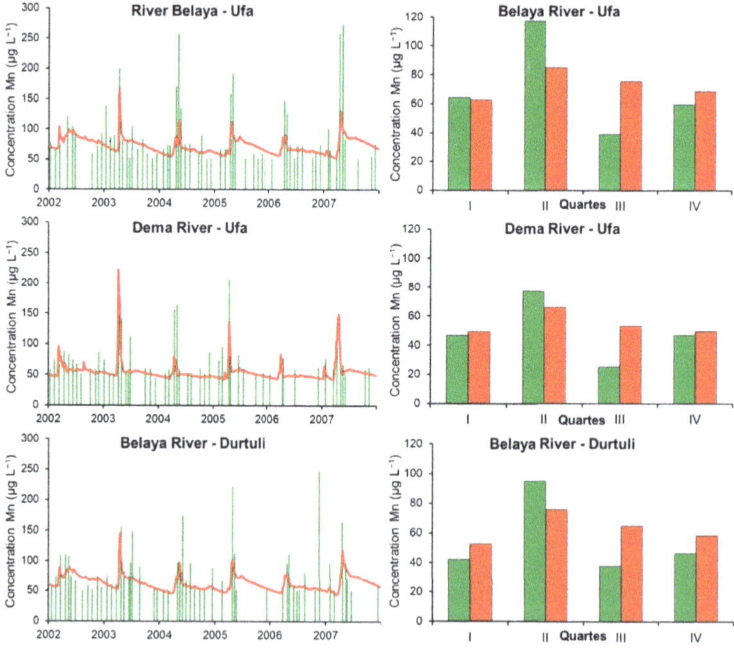

Figure 4. Observed (green) and simulated (red) Mn concentrations at certain hydrochemical monitoring points over the 2002–2007 period: On the left, the dynamics with daily resolution; on the right, the mean intra-annual concentrations by quarters.

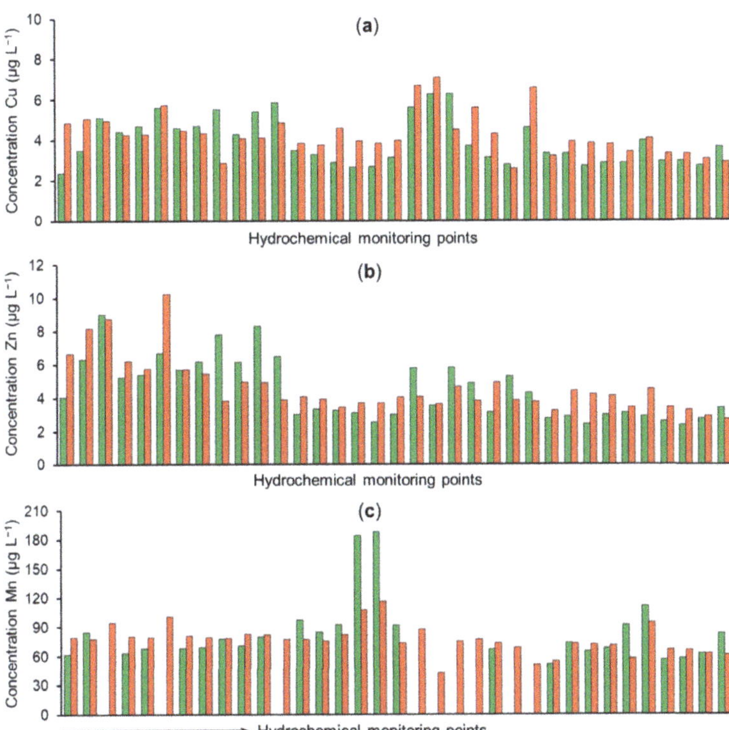

Figure 5. Observed (green) and simulated (red) long-term mean annual heavy metal (HM) concentrations at the hydrochemical monitoring points located in the order from the upper reaches of the Belaya River toward its inflow into the NKR: (**a**) Cu; (**b**) Zn; (**c**) Mn. The arrow shows the direction of river flow.

Model balance calculations show that the total wash-off of metals in the rivers from the NKR catchment averaged over the period of simulation is formed by approximately 80–85% due to leaching from the soil-ground layer. Surface wash-off of metals into rivers usually does not exceed 20%. The share of HM entering the river network with wastewater discharges is small and amounts to ~1% of the total load of copper, ~4% of zinc, and 0.1% of the total load of manganese. A significant proportion of metals washed out from the catchment area accumulate with sediments in the river bottom: On average, ~46% copper, 62% zinc, and 30% manganese. These estimates are in good agreement with the results of HM retention in the Elbe River systems obtained using the METALPOL model [23]: For copper, on average, 51% with a range of variation from 35 to 60% for different tributaries; and for zinc, on average, 45% with a range of variation from 30 to 54%.

An example of the inter-annual dynamics of various sources of river water pollution and a comparison of the actual and simulated Mn load in the NKR is shown in Figure 6. The differences between the compared annual values can be largely explained by significant errors in determining the concentration of heavy metals in river water.

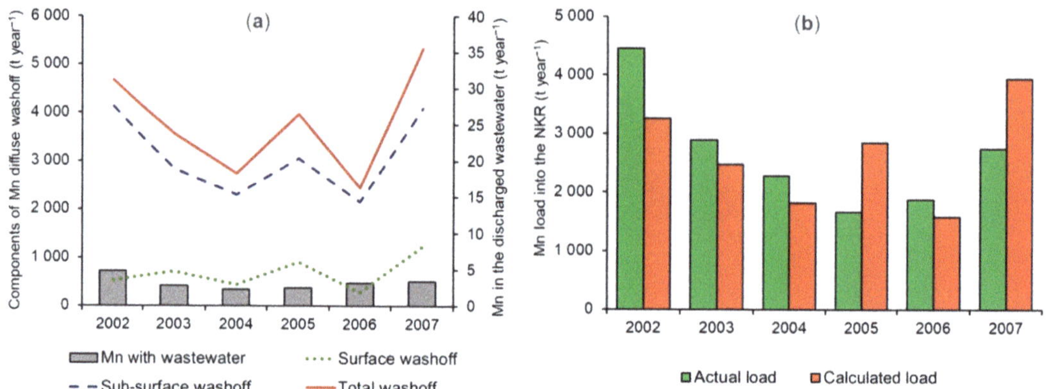

Figure 6. Inter-annual dynamics: (**a**) Simulated components of Mn load in the river network; (**b**) The actual and calculated Mn load in the NKR.

Inter-annual changes in the HM load in the river network and NKR are characterized by the increase in high-water years (2002, 2007), decrease in medium years (2003, 2005), and minimum values in low-water years (2004, 2006). Such inter-annual changes in the HM load occur when HMs enter the river network mainly from diffuse sources from the catchment area, which are affected by climatic factors. Thus, the results presented in this section indicate that, on the whole, the model adequately reproduces the main spatio-temporal patterns of HM runoff formation in the river basin and HM content in the river network.

3.1.2. Mapping of the Simulated HM Characteristics

Regularities of water quality formation in river basins are associated with the study of genetic components of river and chemical runoff; therefore, among the mapped characteristics, we will consider the HM concentrations in river waters and the HM wash-off with various genetic components of river runoff. The algorithm for determining spatial distributions of components of HM runoff based on the developed model was as follows. Daily fields of weather characteristics in the catchment area were constructed using data of meteorological stations, and the ECOMAG-HM model was applied to simulate runoff and HM removal in the local river network by surface, soil, and ground flow for all elementary watersheds. By averaging the daily fields over a multi-year period, maps of the mean annual specific HM load in the river network by various genetic components were obtained for cooper [32] and zinc [34].

The analysis of these maps demonstrates the spatial differences in the predominance of various genetic components in the formation of hydrochemical metal runoff in the NKR basin. Over most of the catchment area, the total HM load into the river network was formed mainly due to subsurface (soil and ground) runoff. The HM wash-off by surface water is almost an order of magnitude smaller, except for the western province of the basin, where the wash-off by surface runoff was comparable or slightly higher than the HM leaching by subsurface runoff.

The map of the mean annual Cu, Zn, and Mn load in the river network shows the close correlation with the spatial distribution of the initial HM content in soils (Figure 7) due to the predominance of HM leaching from the underground zone in the water quality formation.

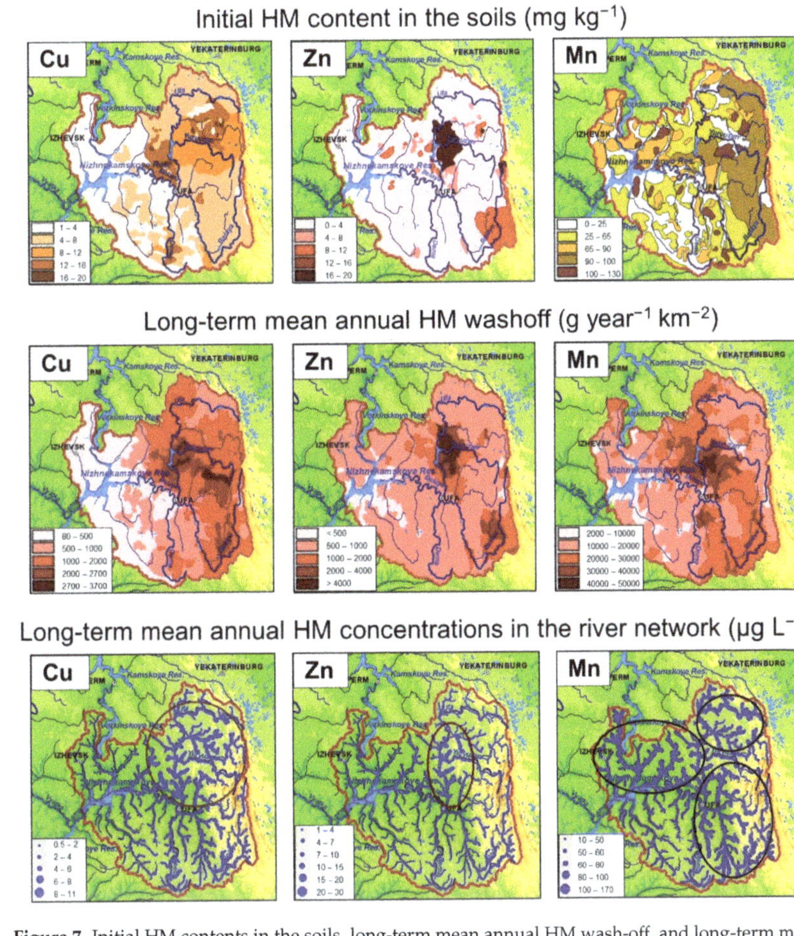

Figure 7. Initial HM contents in the soils, long-term mean annual HM wash-off, and long-term mean annual HM concentrations in the river network.

In turn, the spatial distribution of HM load affects the distribution of HM concentrations in the river network. A map of the simulated mean annual HM concentrations in river water was created (Figure 7) to analyze the spatial distribution of HM content in the NKR drainage basin, including the rivers not covered by hydrochemical observations. This map was obtained by averaging the simulated daily HM concentrations in the elements of the modelled river network over the period of simulation. The line thickness indicates the HM concentration in the river network in accordance with the legend.

As can be seen from Figure 7, small rivers that mainly flow near foci with a high content of HM in soils have higher levels of HM pollution. These areas with increased river pollution are delineated with shafts. Figure 7 clearly shows that a narrow area of increased concentrations of zinc in rivers is formed near the focus of increased zinc content in the soil in the area of the Ufa River catchment. The area with a high copper content in the rivers is noticeably larger and mainly covers the north-eastern part of the NKR basin. The distribution of the increased content of manganese in the soil in the catchment area is spotty; therefore, foci of increased concentrations of manganese in the river water are found everywhere, with the exception of the southwestern part of the basin. The maximum mean annual HM concentrations in these streams exceed the similar average for the other part of the NKR basin by 2–6 times and reach 11 $\mu g \cdot L^{-1}$ for copper, 29 $\mu g \cdot L^{-1}$ for zinc, and 166 $\mu g \cdot L^{-1}$ for manganese. When small, polluted tributaries flow into larger rivers, HM concentrations in river waters decrease as a result of dilution by cleaner water in larger watercourses. A similar effect was obtained in the results of simulating HM concentrations in river waters near polluted areas in the Liuyang River basin in China impacted by mining activities using the SWAT-HM model [31]. As already noted, hydrochemical monitoring points are usually located on large and medium-sized rivers in the immediate vicinity of large settlements and industrial enterprises, so the existing monitoring network does not cover small rivers that are most susceptible to diffuse pollution.

The analysis of simulated results shows that the water quality in the river monitoring points does not meet the MAC standards for fishing use regarding copper (by 2–7 times) and manganese (by 4–19 times). The entire range of changes in zinc concentrations is within the MAC standard. The water quality regarding copper and zinc in all monitoring points is suitable for drinking use. The content of manganese in river water exceeds the MAC standards for drinking use only at several points in some seasons of the year.

3.2. Modeling the Impact of Economic Activity on River Water Pollution

3.2.1. Change in the Amount of Metals Discharged in Wastewater

Intra-annual distribution. Figure 8 shows the simulated dynamics of the copper content in the Belaya River at Ufa city with actual wastewater discharges, as well as with the 60-times increase in copper content in wastewater. At the current level of the copper content in wastewater, its maximum concentration in the river water is simulated during high water periods (snowmelt and rainfall runoff) as a result of the metal load, mainly from diffuse sources. When the multiplicity of the increase in the content of metals in wastewater relative to the existing level exceeds a certain critical value, the following pattern occurs: During periods of high water, the lowest concentrations of metals begin to be observed in contrast to the original case (Figure 8). This occurs as a result of diluting highly polluted industrial wastewater in rivers with significantly less polluted snow and rainwater. Nomograms can be constructed for hydrochemical monitoring stations, showing the "critical" levels of anthropogenic load (the multiplicity of increase in metals discharged from wastewater relative to the existing level), at which such changes occur based on the results of numerical experiments. In particular, in the example shown in Figure 9, the relationship between the river discharges and maximal concentrations of copper in the Belaya River at Ufa city is reversed, when the discharge of copper into the river increases by 20 or more times.

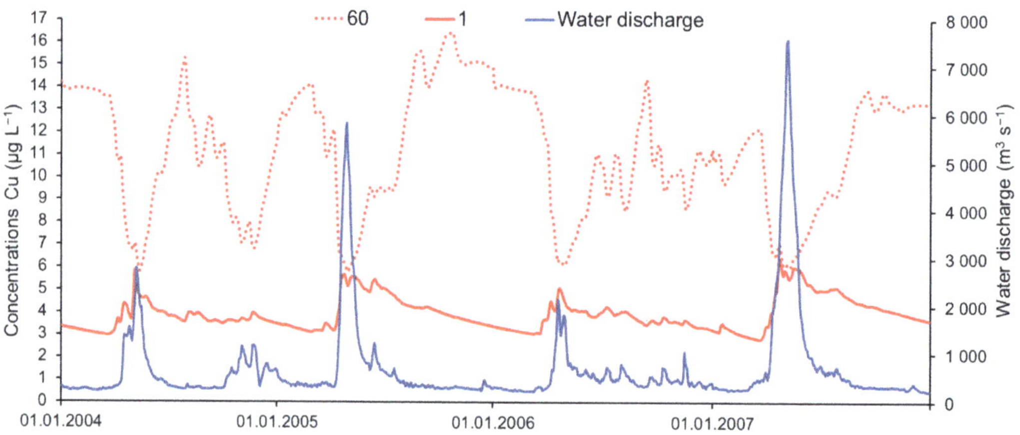

Figure 8. Simulated hydrograph (blue) and dynamics of the copper content in the Belaya River (Ufa city) with the actual content of copper in wastewater (1) and with an increase in the copper content by 60 times (60) over the 2004–2007 period.

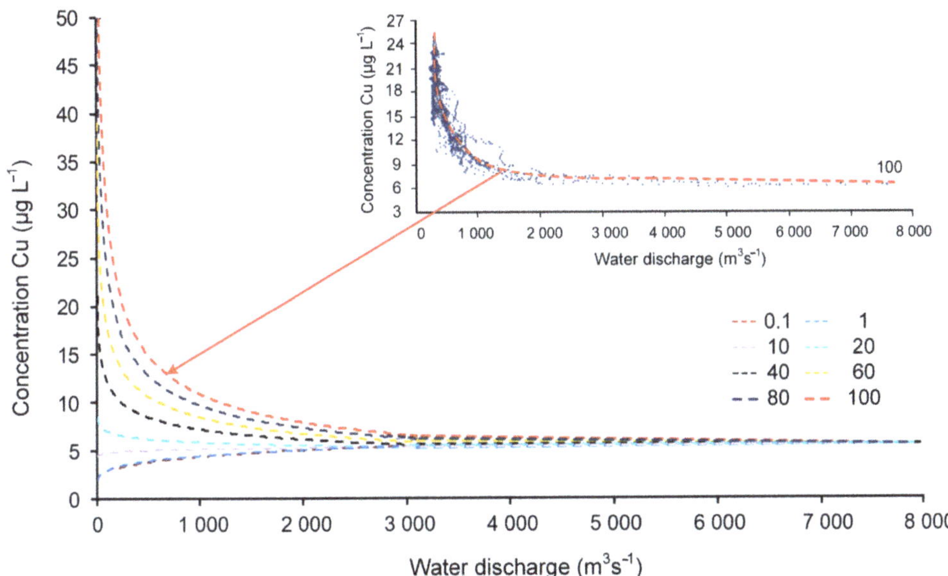

Figure 9. Relationship between water discharges at the Belaya River (Ufa city) and the concentration of copper in river water with the actual content of copper in wastewater (1) and under scenarios of its change at 0.1, 10, 20, 40, 60, 80, 100 times.

Numerical experiments have shown (Figure 9) that with an increase in discharge of the Belaya River near the city of Ufa over 3500 m^3·s^{-1}, copper concentrations in river water asymptotically approach the value of ~6 µg·L^{-1} under various scenarios of wastewater loads. In other words, the contribution of point sources of pollution due to dilution is minimized during periods of high water, and the concentration of metals in river water is mainly determined by the maximum exchange capacity of the catchment leaching of HM from the watershed by surface and subsurface flow during intensive snowmelt or rainfall that forms the maximum runoff. This constant concentration can be named the equilibrium concentration of diffuse saturation of the river basin for this type of HM.

Mean annual characteristics. With an increase in the HM content in wastewater, the contribution of point sources to the river's pollution in the NKR basin increases (Figure 10a). Thus, with a 20-fold increase in HM content in wastewater, the mean annual contribution of point sources to river water pollution is ~20% for copper, 44% for zinc, and 2% for manganese. A 100-fold increase in HM load increases the contribution of point sources to river pollution to ~53%, 80%, and 8%. An increase in the content of HM in wastewater also leads to an increase in their inflow by river water to the reservoir (Figure 10b). Thus, a 100-fold increase in metals causes an increase in the inflow into the NKR of copper by ~2 times, zinc by ~4 times, and manganese by 1.1 times.

3.2.2. Simulation of a Disaster with a Salvo Discharge of Pollutants

Figure 11 shows simulated changes in the maximum concentration of copper along the length of the river under various scenarios of copper salvo discharge (5000, 4000, 3000, 2000, 1000, 500, and 100 kg·day^{-1}) into the Belaya River near Salavat city. As the contamination spot is removed from the site of the salvo discharge, the concentration of copper decreases intensively due to the dilution by the tributaries that flow into the Belaya River, and copper deposits on the river bottom along with sediments. As early as passing Ufa city (~300 km from the accident site), the maximum concentrations of copper are reduced by about two orders of magnitude.

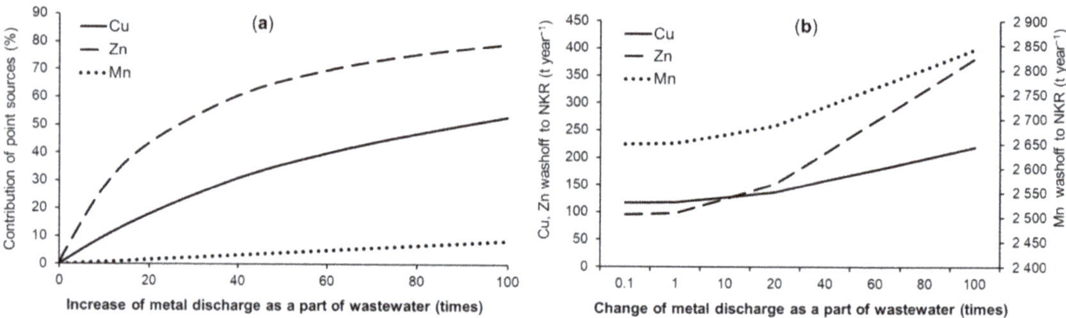

Figure 10. Simulated anthropogenic impact on the NKR basin with changes in the metal content in the wastewater: (**a**) Contribution of point sources to the river water pollution, (**b**) wash-off of metals to the NKR.

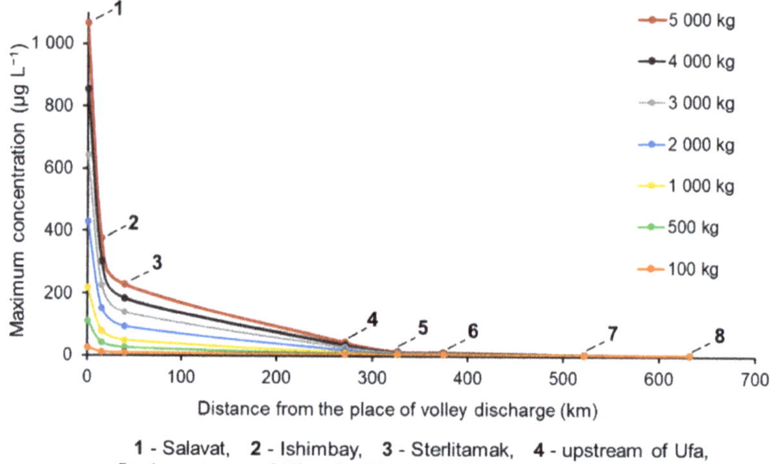

Figure 11. Changes in the maximum concentration of copper at several hydrological stations along the Belaya River under various scenarios of copper dumping in the Salavat city (5000 kg, . . . , 100 kg).

3.2.3. Simulation of Self-Cleaning of the NKR Watershed under Exclusion of Anthropogenic Impact on the Catchment

The estimation of time scales of the catchment's self-purification from HM according to scenario 3 was carried out by simulating the dynamics of HM reserves with the complete exclusion of any anthropogenic impact on the basin over a 400-year period. Based on the maps of metal content in the soil [41], it was determined that in the NKR catchment, the initial reserves of copper are ~1000, zinc ~550, and manganese ~11,350 thousand tons (Figure 12a). The annual flow of copper, averaged over a 33-years period, from the catchment to the river network at the beginning of the calculation period was 147, zinc ~156, and manganese ~1745 t·year^{-1} (Figure 12b). After 200 years, calculations showed a decrease in the removal of copper, zinc, and manganese from the catchment area, respectively, to 140, 147, and 1700 t·year^{-1}. After 400 years, the average annual removal of metals from the river network continued to decline, and amounted to 135, 140 t, and 1640 t year^{-1}, respectively. Thus, in the absence of external impacts, the catchment area was slowly cleared and the content of copper in soils decreased to ~950, zinc to ~500, and manganese to ~ (Figure 12a) thousand tons. As a result of the reduction of metals reserves in the catchment area, their content in the river water decreased. Calculations have shown that over a 400-year period, due to the self-cleaning of the NKR basin, the average annual concentrations of copper, zinc, and manganese in river water decreased slightly by 6–8%.

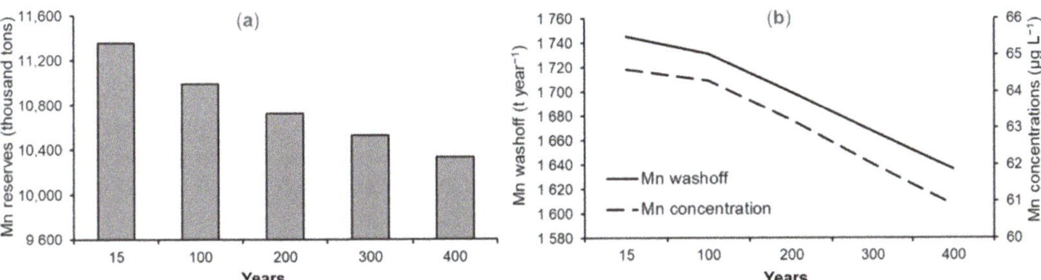

Figure 12. Changes in Mn reserves in soils (**a**), wash-off intensity, and Mn concentration in river waters (**b**) over a 400-year self-purification period.

3.3. Sensitivity of the Water and Chemical Runoff to Climatic Changes

Modeling the water and chemical runoff for the future period, up to 2050, was carried out using a DC climatic scenario, assuming changes in the climate parameters to be the same as the previous observation periods (increments of the daily air temperature of 1 °C and daily precipitation of 10%). Table 4 presents the results of numerical experiments for the simulation of river runoff characteristics under both historical data series and the DC scenario for the 2002–2007 period.

Table 4. Simulated runoff characteristics under historical data series (Hist.) and DC scenario (DC) for the 2002–2007 period.

Years	Annual Runoff (km^3)			Maximal Discharge (m^3·s^{-1})		
	Hist.	DC	Changes (%)	Hist.	DC	Changes (%)
2002	50.3	54.1	8	5503	4750	−14
2003	38.5	40.1	4	3911	4790	22
2004	35.0	39.1	12	3877	3898	1
2005	43.4	45.6	5	8263	8199	−1
2006	32.4	35.2	9	3038	3382	11
2007	55.2	58.6	6		9900	−5
Mean	42.5	45.4	7	5842	5820	−0.4

The mean annual inflow into the NKR under the DC scenario increased by 7% with a range from 4 to 12% in different years compared to the runoff simulated under historical meteorological data. Increments of the maximal discharges of the spring runoff in different years are multidirectional and vary from −14 to 22%. Both increments of runoff and changes in maximal discharges are not related to the amount of annual runoff. They are determined by the conditions of runoff formation during spring in specific years, and in particular, by the factor of "friendly" spring. Let us analyze this in more detail on the example of flow hydrographs for 2002–2003 (Figure 13).

The winter and pre-spring period of 2002 was characterized by an unstable temperature regime with frequent thaws. The increase of 1 °C in the DC scenario caused earlier snowmelt and the formation of a much larger first wave of spring runoff than those simulated from historical meteorological data. The maximum discharge in the second wave of spring runoff decreased by 14% due to the fact that a significant part of the snow water equivalent was consumed during the first wave of spring runoff.

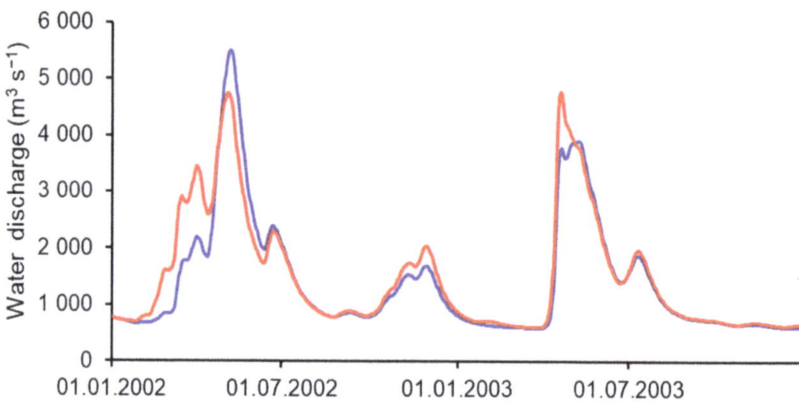

Figure 13. Simulated hydrographs of lateral inflow into the NKR under historical (blue) and DC scenarios (red) over the 2002–2003 period.

The winter and pre-spring season of 2003 was much more severe in terms of temperature conditions than in 2002. Therefore, the increase in temperature by 1 °C did not cause earlier snow melting. The increased maximal snow water equivalent due to both the addition of precipitation in the winter season and more intense snowmelt in the spring under the DC scenario caused a 22% increase in maximal discharge of spring runoff, with the peak 18 days earlier than under the historical scenario.

It is more difficult to generalize the patterns of the influence of climate change on the characteristics of water quality. On the whole, the mean annual Mn concentrations in the last downstream hydrochemical station on the Belaya River (Durtuli gauge) averaged over 7 years in the DC scenario increased by 5% compared to those modeled based on historical meteorological data (Table 5).

Table 5. Simulated water quality characteristics under historical data series and DC scenario for the 2002–2007 period.

Years	Mn Annual Concentrations, Durtuli Gauge ($\mu g \cdot L^{-1}$)			Maximal Concentrations, Durtuli Gauge ($\mu g \cdot L^{-1}$)			Annual Mn Loads into the NKR ($t \cdot year^{-1}$)		
	Hist.	DC	Changes (%)	Hist.	DC	Changes (%)	Hist.	DC	Changes (%)
2002	70.5	72.4	3	88.3	92	4	3257	3553	9
2003	64.5	66.2	3	145	140	−3	2474	2650	7
2004	55.5	59.4	7	96	90	−6	1811	2139	18
2005	61.3	64.8	6	100	92	−8	2845	3032	6
2006	54.8	58.5	7	77	85	−10	1582	1841	16
2007	66.3	71.5	8	116	113	−3	3936	4337	10
Mean	62.2	65.5	5	104	102	−2	2651	2925	10

The changes in annual Mn loads into the NKR in different years are proportional to increments in the mean annual runoff multiplied by the changes in annual concentrations at the last monitoring gauge on the Belaya River. The maximum Mn concentrations averaged over a 7-year period decreased slightly by 2%, and the range of changes in different years vary from 4 to −10%.

Let us compare these values with assessments of the climate change impact on the quantity and quality of water in the Elbe River basin performed using the process-oriented SWIM model [21]. Using climate change scenarios close to the one we employed, river runoff will increase by an average of 11% by the middle of the century. The predicted changes in the concentrations of nitrogen and phosphorus nutrients show multidirectional changes. Uncertainties and high spatial variability of characteristics within the basin may be associated with the schematization of the Elbe basin into a number of separate tributary catchments and with the need to calibrate the model parameters for each of these sub-catchments. In the ECOMAG-HM model, the whole basin and spatial variability of characteristics are simulated with a single set of model parameters.

4. Conclusions

The semi-distributed, physically based ECOMAG-HM model developed for modeling the hydrochemical cycle of heavy metals at the catchment scale was calibrated and verified for the large watershed of the Nizhnekamskoe reservoir according to hydrometeorological, hydrochemical, and water management monitoring data. The model was performed by comparing the observed and simulated river runoff hydrographs and behavior of HM (Cu, Zn, and Mn) concentrations in monitoring points on rivers of the NKR basin. Comparing results showed that seasonal changes in the metal content are characterized by increased concentrations during the high-water period (snowmelt and rainfall runoff). Inter-annual changes in the observed and simulated HM load into the river network and NKR are characterized by its increase in high-water years and minimum values in low-water years. Such seasonal and inter-annual changes in the HM load occur when HMs mainly enter the river network from diffuse sources of pollution on the catchment area caused by climatic factors. Model balance calculations have shown that the mean annual wash-off of metals into the rivers from the NKR catchment is approximately 80–85% attributed to leaching from the soil-ground layer, and the surface wash-off of metals usually does not exceed 20%. The contribution of HMs entering the river network with wastewater discharges is small and does not exceed 4%. A significant part (from 30 to 62%) of metals washed into rivers accumulate with sediments at the river bottom.

The simulated map of the total HM wash-off into the river network closely correlates with the spatial distribution of the initial content of HM in soils due to the predominance of HM leaching from the soil-ground zone in the diffuse sources of river pollution. The initial HM content in soils affects the distribution of HM concentrations in the river network to a large degree. As a result, the most polluted small rivers flow near areas with a high content of HM in the soils. This leads to the important consequences: Without maps of the initial content of heavy metals in soils, it is impossible to simulate spatial differences in water pollution in the river network.

Numerical experiments have shown that with an increase in discharge of the Belaya River more than a certain critical value, the contribution of point sources of pollution is minimized, and the concentration of metals in river water is mainly determined by the maximum exchange capacity of the catchment leaching HM from the watershed by surface and subsurface flows during intensive snowmelt or rainfall that forms the maximum flow; for the city of Ufa, the critical discharge is about 3500 $m^3 \cdot s^{-1}$, above which the copper concentrations in river water asymptotically approach the value of ~6 $\mu g \cdot L^{-1}$ under various scenarios of wastewater loads. This constant concentration can be called the equilibrium concentration of diffuse saturation of the river basin for this type of HM.

The simulation of time scales of the catchment's self-purification from HMs under the scenario with complete exclusion of anthropogenic impacts on the basin has shown that over a 400-year period, the mean annual HM concentrations in the river water decreased by 6–8%. The low rate of reduction of heavy metal concentrations allows us to recommend their current annual values as background values in the NKR basin.

Modeling the river runoff for the future period up to 2050 under a DC climatic scenario showed that the mean annual inflow into the NKR increases by 7% with a range from 4 to 12% in different years. No significant changes in the characteristics of water quality for heavy metals should be expected in the near future. For example, averaged over 7 years, the mean annual Mn concentration increased by 5%, the maximum Mn concentration decreased slightly by 2%, and the annual Mn load into the NKR increased by about 10% compared to those modeled based on historical meteorological data.

The presented results show that the ECOMAG-HM model can be used as an information support tool for assessing water quality characteristics under possible climatic and anthropogenic changes in scientific and applied projects.

Author Contributions: Methodology, writing—review and editing, supervision, funding acquisition, Y.M.; data curation, formal analysis, visualization, project administration, T.F.; conceptualization, validation, investigation, resources, writing—original draft preparation, Y.M. and T.F. All authors have read and agreed to the published version of the manuscript.

Funding: This study was carried out under the Governmental Order to the Water Problems Institute, Russian Academy of Sciences, subject no. 0147-2019-0001.

Institutional Review Board Statement: Not applicable.

Informed Consent Statement: Not applicable.

Data Availability Statement: Not applicable.

Conflicts of Interest: The authors declare no conflict of interest.

References

1. Maistrenko, V.N.; Khamitov, R.Z.; Budnikov, G.K. *Ecological-Analytical Monitoring of Superecotoxicants*; Khimiya: Moscow, Russia, 1996; 319p. (In Russian)
2. Kumar, A.; Pinto, M.C.; Candeias, C.; Dinis, P.A. Baseline maps of potentially toxic elements in the soils of Garhwal Himalayas, India: Assessment of their eco-environmental and human health risks. *Land Degrad. Dev.* **2021**, *32*, 3856–3869. [CrossRef]
3. Moiseenko, T.I.; Morgunov, B.A.; Gashkina, N.A.; Megorskiy, V.V.; Pesiakova, A.A. Ecosystem and human health assessment in relation to aquatic environment pollution by heavy metals: Case study of the Murmansk region, northwest of the Kola Peninsula, Russia. *Environ. Res. Lett.* **2018**, *13*, 065005. [CrossRef]
4. Federal Service for Hydrometeorology and Environmental Monitoring of Russia. *Surface Water Quality in the Russian Federation. Information about the Most Heavily Polluted Water Bodies in the Russian Federation (Appendix to Yearbooks 2011–2019)*; Roshydromet, Gidrokhim. Inst.: Moscow/Rostov-on-Don, Russia, 2020. (In Russian)
5. Ministry of Nature Management and Ecology of the Republic of Bashkortostan. *State Reports on the State of Natural Resources and the Environment in the Republic of Bashkortostan over 2005–2018*; Minecologii RB: Ufa, Russia, 2006. (In Russian)
6. Blokov, I.P. *Environment and Its Protection in Russia. Changes over 25 Years*; Sovet Grinpis: Moscow, Russia, 2018; 432p. (In Russian)
7. Nikanorov, A.M.; Minina, L.I.; Lobchenko, E.E.; Emelyanova, V.P.; Nichiporova, I.P.; Lyampert, N.A.; Pervysheva, O.A.; Lavrenko, N.Y. *Dynamics of Surface Water Quality in Large River Basins in the Russian Federation*; Gidrokhim. Inst.: Rostov-on-Don, Russia, 2015. (In Russian)
8. Ministry of Natural Resources and the Environment of the Russian Federation. *State Report "On the State and Use of Water Resources of the Russian Federation in 2016"*; NIA-Priroda: Moscow, Russia, 2017; 300p. (In Russian)
9. Shcherbakov, B.Y.; Chilikin, A.Y.; Izhevskii, V.S. Volley discharges of industrial sewage and their consequences. *Ekol. Promst. Ross* **2002**, *6*, 39–41. (In Russian)
10. Selezneva, A.V. Anthropogenic load on rivers from point pollution sources. *Izv. Samar. Nauh. Ts. Ross. Akad. Nauk.* **2003**, *5*, 268–277. (In Russian)
11. Bedient, P.B.; Lambert, J.L.; Springer, N.K. Stormwater pollutant load runoff relationships. *J. Water Pollut. Control Fed.* **1980**, *52*, 2396–2404.
12. Knisel, W.G. (Ed.) CREAMS: A field scale model for Chemical, Runoff and Erosion from Agricultural Management Systems. In *USDA Conservation Research Report, No. 26*; Dept. of Agriculture, Science and Education Administration: Washington, DC, USA, 1980; 690p.
13. Gao, L.; Li, D. A review of hydrological/water-quality models. *Front. Agr. Sci. Eng.* **2014**, *1*, 267–276. [CrossRef]
14. Krysanova, V.; Hattermann, F.; Huang, S.; Hesse, C.; Voß, A. Water quality modelling in mesoscale and large river basins. In *Encyclopedia of Life Support Systems, Volume 2: Hydrological Systems Modeling*; Eolss Publishers Co. Ltd.: Oxford, UK, 2009; pp. 11–48.
15. Harmel, R.D.; Smith, P.K.; Migliaccio, K.W.; Chaubey, I.; Douglas-Mankin, K.R.; Benham, B.; Shukla, S.; Muñoz-Carpena, R.; Robson, B.J. Evaluating, interpreting, and communicating performance of hydrologic/water quality models considering intended use: A review and recommendations. *Environ. Model. Softw.* **2014**, *57*, 40–51. [CrossRef]
16. Yang, Y.S.; Wang, L. A review of modelling tools for implementation of the EU Water Framework Directive in handling diffuse water pollution. *Water Resour. Manag.* **2010**, *24*, 1819–1843. [CrossRef]
17. Abbaspour, K.C.; Rouholahnejad, E.; Vaghefi, S.; Srinivasan, R.; Yang, H.; Kløve, B. A continental-scale hydrology and water quality model for Europe: Calibration and uncertainty of a high-resolution large-scale SWAT model. *J. Hydrol.* **2015**, *524*, 733–752. [CrossRef]

18. Santhi, C.; Kannan, N.; Arnold, J.G.; Di Luzio, M. Spatial calibration and temporal validation of flow for regional scale hydrologic modeling. *J. Am. Water Resour. Assoc.* **2008**, *44*, 829–846. [CrossRef]

19. Galván, L.; Olías, M.; de Villarán, R.F.; Domingo Santos, J.M.; Nieto, J.M.; Sarmiento, A.M.; Cánovas, C.R. Application of the SWAT model to an AMD-affected river (Meca River, SW Spain). Estimation of transported pollutant load. *J. Hydrol.* **2009**, *377*, 445–454. [CrossRef]

20. Zhang, P.; Liu, Y.; Yu, Z. Land use pattern optimization based on CLUE-S and SWAT models for agricultural non-point source pollution control. *Math. Comput. Model.* **2013**, *58*, 588–595. [CrossRef]

21. Hesse, C.; Krysanova, V. Modeling climate and management change impacts on water quality and in-stream processes in the Elbe River basin. *Water* **2016**, *8*, 40. [CrossRef]

22. Fuchs, S.; Scherer, U.; Wander, R.; Behrendt, H.; Venohr, M.; Opitz, D.; Hillenbrand, T.; Marscheider-Weidemann, F.; Götz, T. *Calculation of Emissions into Rivers in Germany Using the MONERIS Model Nutrients, Heavy Metals and Polycyclic Aromatic Hydrocarbons*; UBA Technical Report, No. 46; Federal Environment Agency: Dessau-Roßlau, Germany, 2010; 237p.

23. Vink, R.; Peters, S. Modelling point and diffuse heavy metal emissions and loads in the Elbe basin. *Hydrol. Process.* **2003**, *17*, 1307–1328. [CrossRef]

24. Lindström, G.; Pers, C.; Rosberg, J.; Strömqvist, J.; Arheimer, B. Development and testing of the HYPE (Hydrological Predictions for the Environment) model—A water quality model for different spatial scales. *Hydrol. Res.* **2010**, *41*, 295–319. [CrossRef]

25. Arheimer, B.; Olsson, J. *Integration and Coupling of Hydrological Models with Water Quality Models: Applications in Europe*; SMHI: Norrköping, Sweden, 2003; 53p.

26. Borah, D.K.; Bera, M. Watershed-scale hydrologic and nonpoint-source pollution models: Review of applications. *Trans. ASABE* **2004**, *47*, 789–803. [CrossRef]

27. Ambrose, R.B.; Wool, T.A.; Barnwell, T.O. Development of water quality modeling in the United States. *Environ. Eng. Res.* **2009**, *14*, 200–210. [CrossRef]

28. Tang, T.; Strokal, M.; van Vliet, M.T.H.; Seuntjens, P.; Burek, P.; Kroeze, C.; Langan, S.; Wada, Y. Bridging global, basin and local-scale water quality modeling towards enhancing water quality management worldwide. *Curr. Opin. Environ. Sustain.* **2019**, *36*, 39–48. [CrossRef]

29. Bąk, Ł.; Szeląg, B.; Sałata, A.; Studziński, J. Modeling of heavy metal (Ni, Mn, Co, Zn, Cu, Pb, and Fe) and PAH content in stormwater sediments based on weather and physico-geographical characteristics of the Catchment-Data-Mining Approach. *Water* **2019**, *11*, 626. [CrossRef]

30. Motovilov, Y. ECOMAG: A distributed model of runoff formation and pollution transformation in river basins. In *Understanding Freshwater Quality Problems in a Changing World, Proceedings of H04, IAHS–IAPSO–IASPEI Assembly, Gothenburg, Sweden, 22–26 July 2013*; IAHS Publ. 361: Wallingford, UK, 2013; pp. 227–234.

31. Meng, Y.; Zhou, L.; He, S.; Lu, C.; Wu, G.; Ye, W.; Ji, P. A heavy metal module coupled with the SWAT model and its preliminary application in a mine-impacted watershed in China. *Sci. Total Environ.* **2018**, *613–614*, 1207–1219. [CrossRef] [PubMed]

32. Motovilov, Y.G.; Fashchevskaya, T.B. Simulation of spatially-distributed copper pollution in a large river basin using the ECOMAG-HM model. *Hydrol. Sci. J.* **2019**, *64*, 739–756. [CrossRef]

33. Motovilov, Y.G.; Gelfan, A.N. *Runoff Formation Models in Problems of River Basin Hydrology*; Russian Academy of Sciences: Moscow, Russia, 2018; 300p. (In Russian). Available online: https://www.iwp.ru/upload/iblock/a4a/a4ab6902d99c4fa97e5a0da01859042e.pdf (accessed on 7 November 2021).

34. Fashchevskaya, T.B.; Motovilov, Y.G. Modeling water pollution under different scenarios of zinc load on the Nizhnekamskoe Reservoir watershed. *Water Resour.* **2019**, *46*, S69–S80. [CrossRef]

35. Fashchevskaya, T.B.; Motovilov, Y.G. Simulation of Heavy Metal Pollution of Watercourses in the Basin of the Nizhnekamskoe Reservoir. *Water Resour.* **2020**, *47*, 794–809. [CrossRef]

36. *The Republic of Bashkortostan: Brief Ensyclopedia*; Nauchnoe Izd. "Bashkirskaya Entsiklopediya":: Ufa, Russia, 1996; 672p. (In Russian)

37. Khaziev, F.K. *Soils in the Republic of Bashkortostan and Their Fertility Control*; Gilem: Ufa, Russia, 2007; 285p. (In Russian)

38. Fashchevskaya, T.B.; Polianin, V.O.; Fedosova, L.V. Structural Analysis of Water Quality Formation in an Urban Watercourse: Point, Non-Point, Transit, and Natural Components. *Water Resour.* **2018**, *45*, S67–S78. [CrossRef]

39. Chernogayeva, E.M. *Tendencies and Dynamics of the State and Pollution of the Environment in the Russian Federation According to the Data of Long-Term Monitoring for the Last 10 Years. Analytical Review*; Rosgydromet: Moscow, Russia, 2017; 51p. (In Russian)

40. Abdrakhmanov, R.F.; Chalov, Y.N.; Abdrakhmanova, E.P. *Fresh Groundwater in Bashkortostan*; Informreklama: Ufa, Russia, 2007; 184p. (In Russian)

41. Yaparov, I.M. *Atlas of the Republic of Bashkortostan*; Kitap: Ufa, Russia, 2005; 419p. (In Russian)

42. Sauvé, S.; Manna, S.; Turmel, M.-C.; Roy, A.G.; Courchesne, F. Solid-solution partitioning of Cd, Cu, Ni, Pb and Zn in the organic horizons of a forest soil. *Environ. Sci. Technol.* **2003**, *37*, 5191–5196. [CrossRef] [PubMed]

43. Gelfan, A.; Frolova, N.; Magritsky, D.; Kireeva, M.; Grigoriev, V.; Motovilov, Y.; Gusev, E. Climate change impact on annual and maximum runoff of Russian rivers: Diagnosis and projections. *Fund. Appl. Climatol.* **2021**, *7*, 36–79. (In Russian) [CrossRef]

44. *SNIP-2.01.14-83. Construction Standards and Regulations, Determining Design Hydrological Characteristics*; Stroiizdat: Moscow, Russia, 1985; 36p. (In Russian)

45. Moriasi, D.N.; Arnold, J.G.; Van Liew, M.W.; Bingner, R.L.; Harmel, R.D.; Veith, T.L. Model evaluation guidelines for systematic quantification of accuracy in watershed simulations. *Trans. ASABE* **2007**, *50*, 885–900. [CrossRef]

Article

Natural and Anthropogenic Controls of Groundwater Quality in Sri Lanka: Implications for Chronic Kidney Disease of Unknown Etiology (CKDu)

Sen Xu [1], Si-Liang Li [1,2,*], Fujun Yue [1], Charitha Udeshani [1,3] and Rohana Chandrajith [3,4]

[1] Institute of Surface-Earth System Science, School of Earth System Science, Tianjin University, Tianjin 300072, China; xusen@tju.edu.cn (S.X.); fujun_yue@tju.edu.cn (F.Y.); charitha.udeshani@gmail.com (C.U.)

[2] Tianjin Key Laboratory of Earth Critical Zone Science and Sustainable Development in Bohai Rim, School of Earth System Science, Tianjin University, Tianjin 300072, China

[3] Department of Geology, Faculty of Science, University of Peradeniya, Peradeniya 20400, Sri Lanka; rohanac@pdn.ac.lk

[4] China-Sri Lanka Joint Research and Demonstration Center (JRDC), University of Peradeniya, Peradeniya 20400, Sri Lanka

* Correspondence: siliang.li@tju.edu.cn; Tel.: +86-022-8737-0955

Citation: Xu, S.; Li, S.-L.; Yue, F.; Udeshani, C.; Chandrajith, R. Natural and Anthropogenic Controls of Groundwater Quality in Sri Lanka: Implications for Chronic Kidney Disease of Unknown Etiology (CKDu). *Water* **2021**, *13*, 2724. https://doi.org/10.3390/w13192724

Academic Editor: Fernando António Leal Pacheco

Received: 27 August 2021
Accepted: 30 September 2021
Published: 1 October 2021

Publisher's Note: MDPI stays neutral with regard to jurisdictional claims in published maps and institutional affiliations.

Abstract: Poor groundwater quality in household wells is hypothesized as being a potential contributor to chronic kidney disease of unknown etiology (CKDu) in Sri Lanka. However, the influencing factors of groundwater quality in Sri Lanka are rarely investigated at a national scale. Here, the spatial characteristics of groundwater geochemistry in Sri Lanka were described. The relationships of groundwater quality parameters with environmental factors, including lithology, land use, and climatic conditions, were further examined to identify the natural and anthropogenic controlling factors of groundwater quality in Sri Lanka. The results showed that groundwater geochemistry in Sri Lanka exhibited significant spatial heterogeneity. The high concentrations of NO_3^- were found in the districts that have a higher percentage of agricultural lands, especially in the regions in the coastal zone. Higher hardness and fluoride in groundwater were mainly observed in the dry zone. The concentrations of trace elements such as Cd, Pb, Cu, and Cr of all the samples were lower than the World Health Organization guideline values, while some the samples had higher As and Al concentrations above the guideline values. Principal component analysis identified four components that explained 73.2% of the total data variance, and the first component with high loadings of NO_3^-, hardness, As, and Cr suggested the effects of agricultural activities, while other components were primarily attributed to natural sources and processes. Further analyses found that water hardness, fluoride and As concentration had positive correlations with precipitation and negative correlations with air temperature. The concentration of NO_3^- and water hardness were positively correlated with agricultural lands, while As concentration was positively correlated with unconsolidated sediments. The environmental factors can account for 58% of the spatial variation in the overall groundwater geochemistry indicated by the results of redundancy analysis. The groundwater quality data in this study cannot identify whether groundwater quality is related to the occurrence of CKDu. However, these findings identify the coupled controls of lithology, land use, and climate on groundwater quality in Sri Lanka. Future research should be effectively designed to clarify the synergistic effect of different chemical constituents on CKDu.

Keywords: groundwater; water quality; heavy metals; CKDu; dry zone; fluorosis; arsenic

1. Introduction

Chronic kidney disease (CKD) is a very common public health problem that can be observed in many parts of the world and a higher prevalence is reported from many countries including the USA, Australia, and Japan [1,2]. A CKD of unknown etiology (CKDu) has

---

been reported in some countries in recent years in which the etiology is not recognizable [3]. Such a disease with unknown etiology has been reported in certain parts of the world, especially in Africa, Central America, and Asia [4,5]. CKDu is also found in the rural dry zone regions of Sri Lanka, particularly in the North Central Province adjacent regions for more than two decades [6,7]. Due to its remarkable geographical distribution in some specific regions where groundwater is the main drinking water source, it is generally suspected that long-term exposure to various nephrotoxic elements through drinking groundwater is an important risk factor [7–9]. In the rural dry zone regions of Sri Lanka, groundwater is the primary source of water for both drinking and cooking, and CKDu is considered to be related to poor groundwater quality [10–12].

Groundwater is an indispensable limited resource, playing an important role in the social and economic development throughout the world. High temperature and limited precipitation in the rural dry zone regions of Sri Lanka pose challenges in the supply of adequate water to meet daily needs [10]. Groundwater provides domestic water and also contributes irrigation water supply for rural communities in Sri Lanka, since no or less treatment is often required [10,13]. However, the deterioration of groundwater quality and the depletion of water resources due to population growth and excessive use of pesticides and agrochemicals, pose threats to public health [13]. Previous studies found that the prevalence of CKDu is related to the recharge sources of groundwater, and is significantly higher with the groundwater stagnated as well as groundwater recharged from regional flow paths [8,14]. Therefore, it is suggested that the source, recharge mechanism, and flow pattern of groundwater, as well as geological conditions that would cause natural contamination of groundwater, may be the main causative factors for CKDu [14,15].

Anthropogenic activities such as agricultural and sanitation practices, and environmental factors such as lithology, land use, and climatic conditions impact on the groundwater quality in Sri Lanka [16–18]. Proposed factors related to etiology of CKDu in groundwater include but not be limited to fluoride [8,19–24], hardness [8,15,22], major ions [12,20], heavy metals and metalloids [18,25,26], and agrochemical residues [18,27]. Despite groundwater quality related to the occurrence of CKDu in Sri Lanka has been studied extensively, few studies have been carried out to analyze the natural and anthropogenic factors that control the groundwater geochemistry in Sri Lanka. Therefore, the objective of this paper is to investigate the spatial characteristics of groundwater geochemistry, and its relationships with environmental factors, and also to provide a comprehensive review regarding the relationships between groundwater quality and the occurrence of CKDu in the dry zone of Sri Lanka.

2. Materials and Methods

2.1. Study Area

Sri Lanka (6–10° N, 79–82° E) is an island located in the Indian Ocean, just 800 km north of the equator, close to southeastern coast of the Indian subcontinent, and includes 25 administrative districts with a total area of 65,610 km^2. The length of the island is 440 km from north to south with a maximum width of 226 km. Although the area of the island is relatively small, it has a widely changing topography, geology, and climate within a short distance imparting some unique environmental features (Figure 1) [28]. The topographical configuration of the island is a highland region located in the center, surrounded by a vast lowland plain. Over 90% of the island is underlain by Precambrian rocks. According to a global lithological map [29], metamorphic rocks cover about 68.2% of the island, followed by acid plutonic rocks (14.4%) and unconsolidated sediments (12%). As per the year 2017, the percentages of forest, cropland, and impervious surface of the island were 60.4%, 30.9%, and 0.6%, respectively. The island is situated in the tropical, Indian ocean monsoonal climate region that contains three climate zones based on the amount of precipitation, named as the wet, dry and intermediate zones [13,16]. The western slopes of the central highlands have the highest intensity of precipitation that exceeds 4000 mm per annum according to the records of the average for years 1970–2000 [30]. The lowest precipitation is recorded in the Mannar Island with values about 900 mm per annum. CKDu is found particularly in the dry zone, metamorphic terrain that occupies two-thirds of the country with the annual rainfall of about 1250 mm, while air temperature varies from 28 to 30 °C, with humidity of around 70% [16].

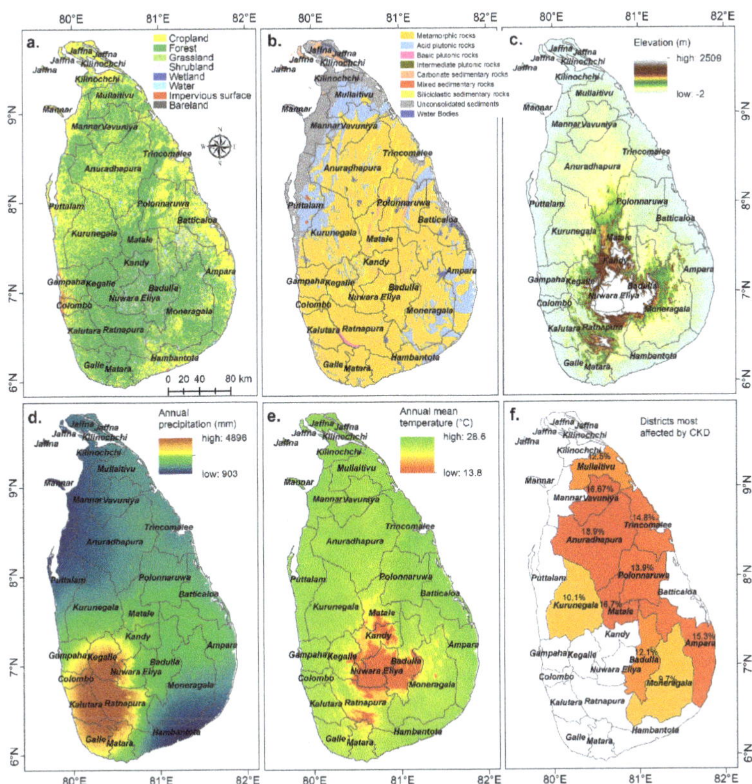

Figure 1. Maps showing (**a**) land use, (**b**) lithology, (**c**) elevation, (**d**) annual precipitation, (**e**) annual mean temperature, and (**f**) CKD prevalence rates across the most affected districts in Sri Lanka [10].

2.2. Data Source

Groundwater quality data including the concentrations of NO_3^-, hardness, fluoride, heavy metals and metalloids (As, Cd, Fe, Mn, Cu, Cr, Pb, and Al) of groundwater samples collected from the wells in Sri Lanka were obtained from Groundwater Quality Atlas of Sri Lanka [31]. A total of 1304 samples were collected and analyzed from 2010 to 2014 for their study. The maximum and average concentrations of the water quality parameters of all the samples in each district were reported by Kawakami et al. (2014) [31], and used and reanalyzed in this study (Table S1).

Lithology, land use, and climatic conditions were quantitatively described for each district to examine the effects of environmental factors on groundwater quality (Table S2). The elevation data was obtained from EarthEnv-DEM90, that is a digital elevation model derived from CGIAR-CSI SRTM v4.1 and ASTER GDEM v2 data products [32]. To quantify the lithological distribution, a global lithological map was used and analyzed for Sri Lanka [29]. Land use dataset with a 10 m resolution was also used to obtain information for this study [33]. Land use types include cropland, forest, grassland, shrubland, wetland, water body, impervious surface, and bare land, among which cropland and impervious surface were mainly focused. Meteorological data including air temperature and precipitation data used in this study were obtained from WorldClim version 2.1 at a spatial resolution of 30 arc-seconds and are the averages for years 1970–2000 [30].

2.3. Statistical Methods

Correlation Analysis (CA) and Principal Component Analysis (PCA) were used to examine the relationships between the groundwater quality parameters for exploring their possible sources. The suitability of data for PCA was checked by Kaiser–Meyer–Olkin (KMO) and Bartlett's sphericity tests. The average concentrations of groundwater quality parameters were used in the CA and PCA. Redundancy analysis (RDA) was selected to assess the explanation of variations of groundwater quality parameters by environmental factors. In the ordination diagram resulted from RDA, the arrows pointing in the same direction represent positive correlations, or vice versa. The data shown in Figure 2 were used in the RDA. The CA and PCA were performed through the statistical software package SPSS 22.0, and the RDA was carried out using CANOCO 5.0 software (Microcomputer Power Company, Ithaca, NY, USA).

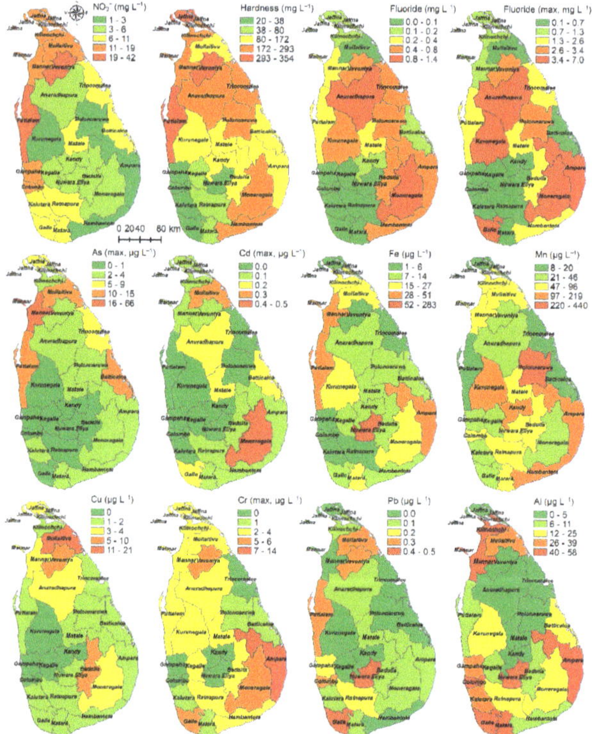

Figure 2. Maps showing the spatial patterns of average concentrations of groundwater quality parameters in each district (maximum concentration (max) was showed for fluoride, As, and Cd). The ranges of data values were categorized into five classes based on the natural break method.

3. Results and Discussion

3.1. Spatial Characteristics of Groundwater Geochemistry

3.1.1. Nitrate, Hardness, and Fluoride

Groundwater quality parameters varied considerably in different regions (Figure 2). The average concentration of NO_3^- in the district ranged from 1 mg/L to 42 mg/L. The highest concentration (366 mg/L) and highest average concentration (42 ± 90 mg/L) of

NO$_3^-$ were observed in the Puttalam district, where 49.7% of the land area is used for agriculture. The World Health Organization (WHO) guideline for NO$_3^-$ in drinking water is 50 mg/L [34]. The maximum NO$_3^-$ concentration in 13 out of 25 districts exceeded 50 mg/L limit. Higher concentrations of NO$_3^-$ were also found in other districts where a high percentage of land is used for agricultural purposes, for instance, the Jaffna (62.9%), Mannar (35.3%), and Vavuniya (28%) districts. Agricultural regions in the coastal zone are mainly underlain by recent unconsolidated sediments in which farmers excessively use a large amount of nitrate fertilizers for their cultivations, leading to nitrate pollution of groundwater [35]. Manure may be a cause for the high NO$_3^-$ concentration found in the Vavuniya district [36].

The average hardness ranged from 20 to 354 mg/L. The highest hardness (1734 mg/L) was found in the Hambantota district, while the highest average hardness (354 ± 190 mg/L) was noted in the Jaffna district, where the percentage of carbonate sedimentary rocks (37.1%) is the highest. The Puttalam and Vavuniya districts also showed a higher hardness. Since water hardness poses no apparent health concerns, there are no guidelines or regulations for optimum hardness in drinking water [34]. Considering the impact of hardness on water palatability, a non-health-based standard of 250 mg/L (as CaCO$_3$) was defined by the Sri Lankan Standard for drinking water [37]. The geographic distribution of CKDu cases in the endemic region showed a strong association with the consumption of hard water [27].

The average concentration of fluoride ranged from <0.02 mg/L to 1.4 mg/L in Sri Lanka. The highest average concentration of fluoride (1.4 ± 1.3 mg/L) was recorded in the Moneragala district. High concentrations of fluoride (max: 7 mg/L; ave: 1.1 ± 0.9 mg/L) were also found in the Anuradhapura district. Spatially, fluoride showed a high concentration in the dry zone and a low concentration in the wet zone and also in the northern coastal zone. The spatial distribution of fluoride is closely resembled with that of the hardness. Although the maximum guideline value for fluoride in drinking groundwater as suggested by the WHO is 1.5 mg/L [34], and the limit suggested by the Sri Lankan Standard is 1.0 mg/L [37], both values cannot be applicable to the dry zone regions of Sri Lanka due to a higher water consumption for drinking under prevailing hot and dry ambient conditions [19]. The average concentrations of fluoride recorded in all the districts were lower than 1.5 mg/L, and only 9.9% of the samples had a concentration higher than 1.5 mg/L [36]. The fluoride concentration in Sri Lanka is much lower than that in many other regions in the world, however dental and skeletal fluorosis cases are common in these areas [10,16,19].

3.1.2. Heavy Metals and Metalloids

In some studies of potential nephrotoxic effects of environmental exposure to heavy metals and metalloids, mainly As, Cd, and Pb have been identified as causative factors for CKDu [18,25,27,38,39], although some studies reject this hypothesis [7,8,15,20,40]. These metals in groundwater may be derived from rock weathering or by human activities, such as the use of agricultural chemicals [18]. Arsenic in groundwater has been proposed as a causal factor for CKDu in Sri Lanka [18,27]. The WHO guideline value and Sri Lankan standard limit for As are 10 µg/L [34,37]. The highest concentration (66 µg/L) and highest average concentration (7 ± 11.7 µg/L) of As were recorded in the Mannar district. High As concentrations were also found in the Batticaloa, Mullaitivu, and Puttalam districts that had the maximum As concentration higher than 10 µg/L. Some recent detail studies also noted higher As concentrations in groundwater, particularly in groundwater extracted from sedimentary aquifer systems [35,41].

Cd is known to be a toxic metal and has been listed as one of the causes of CKDu due to the well-known nephrotoxicity [12,25]. The highest Cd concentration in this study was 0.5 µg/L and recorded in the Moneragala district. All of the well water samples had a Cd concentration much lower than the WHO guideline value of 3 µg/L [34]. The highest Pb concentration (288 µg/L) was noted in the southern district of Galle, however, it may be an outlier. If this data was eliminated, the highest Pb concentration of 4.6 µg/L was recorded in the Anuradhapura district, but it is lower than the WHO guideline value of 10 µg/L [34].

Fe is one of the most abundant metals in the Earth's crust and is an essential human micronutrient [34]. Mn usually coexists with Fe and is also one of the most abundant metals in the Earth's crust. Cu is an essential nutrient and also a contaminant of drinking water. The highest concentrations and highest average concentrations of Fe (max: 828 µg/L; ave: 283 µg/L), Mn (max: 9772 µg/L; ave: 440 µg/L), and Cu (max: 443 µg/L; ave: 21 µg/L) were recorded in the Nuwara Eliya, Polonnaruwa, and Mullaitivu districts, respectively. The WHO guideline value for Cu in drinking water is 2 mg/L, but no WHO guideline values for Fe and Mn in drinking water are proposed [34]. The highest Cr concentration (14 µg/L) was found in the Ampara district, yet the level is lower than the WHO guideline value for Cr in drinking water of 50 µg/L [34]. The highest concentration (1457 µg/L) and highest average concentration (58 µg/L) of Al were found in the Moneragala and Nuwara Eliya districts, respectively. The maximum guideline value for Al in drinking water suggested by the WHO is 200 µg/L [34], and 2.3% of the samples exceeded the guideline value [36].

3.2. Natural and Anthropogenic Controls on Groundwater Geochemistry

The correlation matrix demonstrated that there were significant positive correlations among NO_3^-, As, and Cr ($p < 0.05$) (Figure 3), indicating the higher contamination of As and Cr of groundwater in agricultural areas. Water hardness was positively correlated with fluoride (R = 0.57, $p < 0.01$), which suggests that they may be affected by similar geogenic factors and processes, such as the weathering of fluoride-bearing minerals under the hot climatic condition and excessive evaporation in the dry zone [19,42]. Cd and Cu showed a significant positive correlation (R = 0.54, $p < 0.01$), implying that they may be derived from the same source.

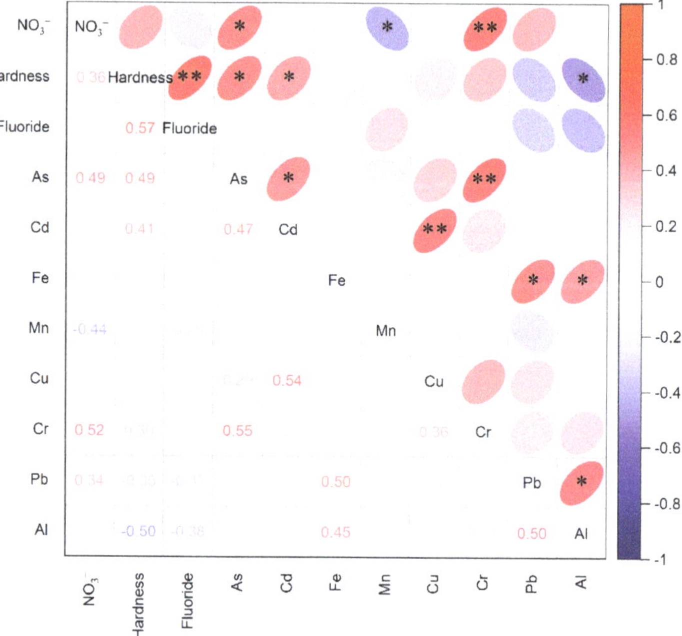

Figure 3. Correlation coefficients for Pearson correlation analysis among groundwater quality parameters. * Correlation is significant at $p < 0.05$; ** Correlation is significant at $p < 0.01$.

PCA was employed to identify the associations between groundwater quality parameters and their potential sources [43]. Four principal components (PCs) with eigenvalues exceeding 1 were identified, and the PCA results are shown in Table 1. The first PC was dominated by NO_3^-, water hardness, As, and Cr, with high loadings of 0.86, 0.67, 0.7, and 0.71, respectively, explaining 26.5% of the total data variance. The second PC explained 24.6% of the total data variance and was mainly contributed by Fe, Pb, and Al, with higher loadings of 0.81, 0.77 and 0.78, respectively. The third PC was mainly contributed by Cd and Cu, with high loadings of 0.88 and 0.76, respectively, explaining 12.6% of the total data variance. The fourth PC was mainly contributed by fluoride and Mn, with high loadings of 0.75 and 0.72, respectively, and explained 9.5% of the total data variance. The four PCs totally accounted for 73.2% of the total data variance. Agrochemicals especially phosphate fertilizer used in agriculture have been identified as the main source of As in areas affected with CKDu in Sri Lanka [18]. The groundwater in the dry zone has high fluoride concentrations due to the water–rock interaction with the aquifer [15,19]. In addition, the occurrence of CKDu is strongly associated with the consumption of hard water in the rural dry zone [22,27]. Therefore, the first PC may be related to the agricultural activities in the rural dry zone, suggesting the contributions of anthropogenic sources to NO_3^-, As, and Cr. In contrast, other PCs may be dominated by natural geogenic sources and processes, especially bedrock weathering.

Table 1. Component matrix and explained variance (principal component analysis (PCA)) for groundwater quality parameters.

Variable	PC1	PC2	PC3	PC4
NO_3^-	0.86	0.03	−0.12	−0.36
Hardness	0.67	−0.47	0.16	0.46
Fluoride	0.08	−0.37	0.04	0.75
As	0.70	−0.01	0.42	−0.05
Cd	0.17	−0.16	0.88	0.05
Fe	0.07	0.81	−0.15	0.24
Mn	−0.34	0.05	−0.01	0.71
Cu	0.20	0.12	0.76	0.01
Cr	0.71	0.24	0.28	0.02
Pb	0.17	0.77	−0.06	−0.27
Al	−0.12	0.78	0.25	−0.22
Eigenvalue	2.92	2.70	1.39	1.05
Variance (%)	26.5	24.6	12.6	9.5
Cumulative (%)	26.5	51.1	63.7	73.2

The groundwater quality in Sri Lanka is mainly influenced by environmental factors such as lithology, land use, and climatic condition, and anthropogenic activities including agricultural practices, disposal of domestic sewage, and industrial effluents [16–18]. To explore the potential influencing factors of groundwater quality in Sri Lanka, the relationships between the environmental factors and the groundwater quality parameters were examined. As shown in Figure 4, water hardness showed a negative correlation with the precipitation (for average hardness: $R^2 = 0.78$, $p < 0.01$; for maximum hardness: $R^2 = 0.53$, $p < 0.01$), following a power law relationship. Similar relationships were also found between fluoride and precipitation, and between As and precipitation. This suggests that in the dry zone with low precipitation, the concentrations of fluoride and As in groundwater were higher than those in the wet zone where excessive precipitation is characterized. Since lower precipitation is generally accompanied by high ambient temperature in Sri Lanka, the higher concentrations of hardness, fluoride, and As were also found in areas with high air temperature.

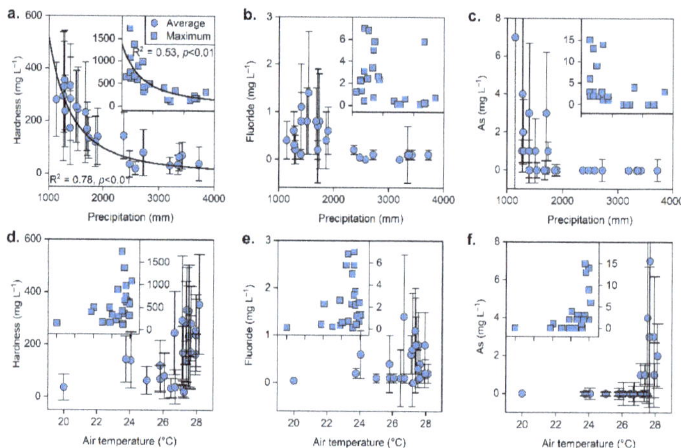

Figure 4. The relationships between hardness, fluoride, and As concentrations, and precipitation (**a–c**) and air temperature (**d–f**).

The positive correlations between NO_3^- and hardness and cropland suggest the effects of agricultural activities on groundwater quality and that most high nitrate regions are underlain by sedimentary limestone sequences (Figure 5). In Sri Lanka, all wells with a high As concentration were located on a specific soil type of "sandy regosols on recent beach and dune sands" [36]. The positive correlation between As concentration and unconsolidated sediments and the spatial characteristics of As concentration indicate that As had a geological source. Amarathunga, et al. (2019) [44] noted that As is released from coatings of sand grains due to reductive dissolution under near natural pH condition in the Mannar region where a majority of wells exceeded the WHO recommended level of As.

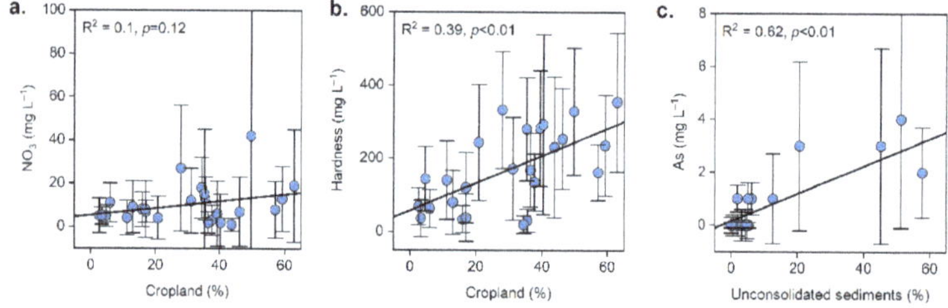

Figure 5. The relationships between NO_3^- concentration (**a**) and hardness (**b**) and cropland, and between As concentration and unconsolidated sediments (**c**).

RDA was used to assess the explanation of variations of the groundwater quality parameters by the environmental factors. The RDA results provided overall descriptions of the influences of environmental factors on groundwater geochemistry [45]. The results showed that the environmental factors can account for 58% of the spatial variation in the overall groundwater geochemistry (pseudo-F = 2.5; p = 0.002). The first two axes explained most of the spatial variation of groundwater geochemistry, with the first axis explaining 26% of the variation and the second axis explaining 16% of the variation. Ordination diagrams resulted from RDA showed the relationships between the groundwater quality parameters and the potential influencing factors (Figure 6). The results indicated that hardness was positively correlated with agricultural lands while fluoride concentration was positively correlated with ambient temperature. By contrast, precipitation and forest were negatively correlated with most of the groundwater quality parameters. These results indicate the synergetic controls of lithology, land use, and climate on the groundwater quality in Sri Lanka.

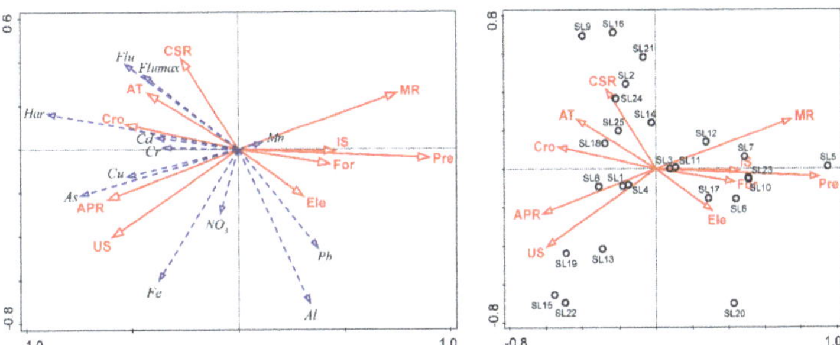

Figure 6. Ordination diagrams showing the relationships between the groundwater quality parameters (represented by blue lines) and the potential influencing factors (represented by red lines) according to the redundancy analysis (RDA). The analyzed potential influencing factors include elevation (Ele), cropland (Cro), forest (For), impervious surface (IS), metamorphic rocks (MR), acid plutonic rocks (APR), unconsolidated sediments (US), carbonate sedimentary rocks (CSR), precipitation (Pre), and air temperature (AT).

3.3. Relationships between Groundwater Quality and Occurrence of CKDu

In the past three decades, possible etiology of CKDu has been extensively explored through multi-disciplinary studies. The CKDu distribution related to geography and socio-economy showed that the possible etiology is related to geogenic environment and occupational factors [6,9,39,46,47]. The long-term exposure to various nephrotoxic elements through drinking groundwater is widely suspected due to the geographical distribution of CKDu [7,8,15,16,20,21,28,40]. CKDu in Sri Lanka was first reported from the north central province of the rural dry zone, and the main victims of CKDu are young male farmers with substandard socio-economic background [6,10,39]. In the worst affected areas of CKDu, more than 98% of the population had completely relied on groundwater as their primary source for drinking or cooking at least five consecutive years between 1999 and 2018 [10]. This indicates that groundwater quality may interfere with human health in CKDu endemic region.

3.3.1. Relationship between Fluoride and CKDu

Fluoride has been proposed as a causal factor for CKDu through drinking water. It is suggested that the source of F is bedrock weathering [19,21,22]. Climatic and hydrological conditions were found to be related to the anomalous concentrations of fluoride in the groundwater in Sri Lanka [19]. Fluoride may be an essential element for humans, for instance, fluoride is a beneficial element in the prevention of dental caries [34]. However, elevated fluoride intakes can have more detrimental effects such as dental and skeletal tissues [34]. Previous studies have revealed the dose–effect relationships between drinking water fluoride levels and damage to kidney functions [11,24]. High fluoride concentrations in drinking water elevated the levels of renal and liver function enzymes in serum and caused severe histological changes of the liver and kidneys [48]. Due to its high rank in the Hofmeister Series for denaturing proteins of the kidney membrane, fluoride may contribute to CKDu [20]. In this study, high fluoride concentrations were found in the dry zone where CKDu has its higher prevalence (Figure 1). Interestingly, no CKDu was observed in the areas in the southeast where the fluoride concentration in groundwater was greater than 1.0 mg/L [31]. This suggests that the high concentration of fluoride in groundwater is not only the factor affecting the occurrence of CKDu. The interaction of fluoride with aluminum to form nephrotoxic aluminum fluoride complexes has been proposed as a cause of CKDu [26]. In addition, when fluoride interacts with major cations

(Ca^{2+}, Mg^{2+}, and Na^+) and metals such as cadmium, it is suspected that it will aggravate kidney failure [12,21,22,28,49]. Therefore, the synergistic effect between fluoride and major ions, and metals on kidney functions should be paid more attention.

3.3.2. Water Hardness and Major Ions

Hardness in water is caused by a variety of dissolved polyvalent metallic ions, predominantly the cations of Ca and Mg. As classification of hardness, the ranges of 0–60 mg/L, 61–120 mg/L, 121–180 mg/L, and >181 mg/L are classified as soft, moderately hard, hard, and very hard, respectively [34]. Previous studies revealed the highly statistically positive correlation between the occurrence of CKDu and the hard water consumption in Sri Lanka, and found that 96% of CKDu patients consumed hard or very hard water from wells in shallow regolith aquifers for at least 5 years [12,27]. However, many studies found that hard to very hard groundwater is not exclusive to all CKDu endemic areas [14,16]. For instance, only 14% of the samples from Mahiyanganaya, a high CKDu endemic dry zone area of Sri Lanka, had hard to very hard groundwater [16]. Moreover, the groundwaters from some areas of Sri Lanka that are not seriously affected by CKDu had very hard waters, such as the districts of Puttalam (329 ± 174 mg/L) and Jaffna (354 ± 190 mg/L). Nevertheless, hardness might interfere with other chemical constituents in water to form various complexes harmful to human health [8,15].

3.3.3. Nephrotoxic Heavy Metals and Metalloids

Heavy metals and metalloids such as As and Cd have been considered as important risk factors for CKDu [18,25,38]. The application of pesticides and fertilizers in rice paddy cultivation in Sri Lanka is a possible source of high Cd [25]. In this study, the Cd concentrations of all the well water samples were significantly below the WHO guideline value. Similar results were also reported by many other studies [16,40,46]. Moreover, previous studies noted that the levels of Cd and As of drinking water and rice samples collected from the affected areas of CKDu were within the levels recommended by WHO and Sri Lankan standard and did not indicate contamination in any form [40]. A number of studies also found that in the CKDu endemic areas of Sri Lanka, the As concentration in groundwater is significantly low [8,15,16,46]. Therefore, it is suggested that As and Cd in groundwater are unlikely to be risk factors for CKDu in Sri Lanka. In addition, Jayasumana et al. (2014) [27] have demonstrated the link between hardness and As. They proposed that As mainly derived from chemical fertilizers and pesticides can eventually damage kidney tissue when combined with Ca and/or Mg in groundwater. In addition, there is considerable evidence that agricultural workers in the CKDu endemic areas are exposed to As, but the exact source and entry mode of As are still controversial [27].

4. Conclusions

This study investigated the spatial characteristics of groundwater geochemistry in Sri Lanka. Groundwater quality data of 1304 water samples collected from 2010 to 2014 were statistically analyzed. Land use and lithology were quantitatively described to examine their relationships with the groundwater quality parameters. The data of ambient temperature and precipitation were used to explore the effects of climatic conditions on groundwater quality. The results showed that the high concentrations of NO_3^- were found in the districts with a high percentage of cropland, especially in the districts in the coastal zone. The Puttalam district had the highest concentration and highest average concentration of NO_3^-. The samples from the dry zone had a higher hardness, and in the Jaffna district, where the percentage (37.1%) of carbonate sedimentary rocks is the highest, the average hardness (354 ± 190 mg/L) of samples was the highest. Similar to the spatial distribution of hardness, fluoride also showed a high concentration in the dry zone, and 9.9% of the samples had a concentration higher than the WHO guideline value of 1.5 mg/L. Heavy metals and metalloids such as As, Cd and Pb have been suggested as causal factors for CKDu in Sri Lanka. The maximum As concentrations of the samples in the Mannar,

Batticaloa, Mullaitivu, and Puttalam districts were higher than the WHO guideline and Sri Lankan standard value (10 µg/L). The Cd and Pb concentrations of all the samples were lower than the WHO guideline values of 3 µg/L and 10 µg/L, respectively. In addition, all of the samples had Cu and Cr concentrations below the WHO guideline values, and 2.3% of the samples had a concentration of Al above the WHO guideline value (200 µg/L).

Significant positive correlations ($p < 0.05$) among NO_3^-, As, and Cr were observed, suggesting the effects of agricultural activities on As and Cr concentrations. The significant positive correlation between water hardness and fluoride indicated that they might be affected by rock weathering in the dry zone. According to the results of PCA, four PCs were identified and explained 73.2% of the total data variance. Considering the concentrations and spatial characteristics of groundwater quality parameters, PC1 with high loadings of NO_3^-, hardness, As, and Cr was considered to be affected by agricultural activities, and other PCs was primarily attributed to natural sources and processes. Hardness, fluoride, and As concentration were positively correlated with precipitation and negatively correlated with air temperature, showing a clear difference in groundwater quality between different climatic regions, and suggesting that groundwater geochemistry in Sri Lanka is closely related to climatic conditions. In addition, NO_3^- concentration and water hardness had positive correlations with cropland, and As concentration had a positive correlation with unconsolidated sediments, implying the effects of land use and lithology on groundwater geochemistry. The RDA results showed that 58% of the spatial variation in the overall groundwater geochemistry can be accounted by the environmental factors.

To our knowledge, there are few previous studies that quantitatively describe land use and lithology, and identify the natural and anthropogenic controlling factors of groundwater quality in Sri Lanka, especially at a national scale. This study provided key insights of the effects of environmental factors on groundwater quality in Sri Lanka, which considerably helps to investigating the relationships between groundwater quality and the occurrence of CKDu. In the future, high spatial-temporal resolution sampling is needed to unravel controlling factors of groundwater quality, and multidisciplinary studies are required to identify risk factors of CKDu in Sri Lanka.

Supplementary Materials: The following are available online at https://www.mdpi.com/article/10.3390/w13192724/s1, Table S1: Statistics of the groundwater quality parameters for each district in Sri Lanka, Table S2: Characteristics of the 25 administrative districts in Sri Lanka.

Author Contributions: Conceptualization, S.X. and S.-L.L.; methodology, S.X. and F.Y.; resources, S.-L.L. and F.Y.; data curation, S.X. and S.-L.L.; writing—original draft preparation, S.X., S.-L.L., F.Y., C.U. and R.C.; writing—review and editing, S.X., S.-L.L., F.Y., C.U. and R.C.; visualization, S.X. and S.-L.L.; supervision, C.U. and R.C.; project administration, S.-L.L. and R.C.; funding acquisition, S.-L.L. and R.C. All authors have read and agreed to the published version of the manuscript.

Funding: This work was carried out as collaborative research, funded by the National Natural Science Foundation, China (Grant No. 41861144026) and National Science Foundation, Sri Lanka (Grant No. ICRP/NSF-NSFC/2019/BS/02).

Institutional Review Board Statement: Not applicable.

Informed Consent Statement: Not applicable.

Data Availability Statement: The data presented in this study are available on request from the corresponding author.

Conflicts of Interest: The authors declare no conflict of interest.

References

1. Atkins, R.C. The epidemiology of chronic kidney disease. *Kidney Int.* **2005**, *67*, S14–S18. [CrossRef]
2. Jha, V.; Garcia, G.G.; Iseki, K.; Li, Z.; Naicker, S.; Plattner, B.; Saran, R.; Wang, A.Y.-M.; Yang, C.-W. Chronic kidney disease: Global dimension and perspectives. *Lancet* **2013**, *382*, 260–272. [CrossRef]
3. Weaver, V.M.; Fadrowski, J.J.; Jaar, B.G. Global dimensions of chronic kidney disease of unknown etiology (CKDu): A modern era environmental and/or occupational nephropathy? *BMC Nephrol.* **2015**, *16*, 145. [CrossRef]

4. Almaguer, M.; Herrera, R.; Orantes, C.M. Chronic kidney disease of unknown etiology in agricultural communities. *MEDICC Rev.* **2014**, *16*, 9–15. [CrossRef]

5. Gifford, F.J.; Gifford, R.M.; Eddleston, M.; Dhaun, N. Endemic nephropathy around the world. *Kidney Int. Rep.* **2017**, *2*, 282–292. [CrossRef] [PubMed]

6. Athuraliya, N.T.; Abeysekera, T.D.; Amerasinghe, P.H.; Kumarasiri, R.; Bandara, P.; Karunaratne, U.; Milton, A.H.; Jones, A.L. Uncertain etiologies of proteinuric-chronic kidney disease in rural Sri Lanka. *Kidney Int.* **2011**, *80*, 1212–1221. [CrossRef] [PubMed]

7. Chandrajith, R.; Nanayakkara, S.; Itai, K.; Aturaliya, T.N.C.; Dissanayake, C.B.; Abeysekera, T.; Harada, K.; Watanabe, T.; Koizumi, A. Chronic kidney diseases of uncertain etiology (CKDue) in Sri Lanka: Geographic distribution and environmental implications. *Environ. Geochem. Health* **2011**, *33*, 267–278. [CrossRef] [PubMed]

8. Balasooriya, S.; Munasinghe, H.; Herath, A.T.; Diyabalanage, S.; Ileperuma, O.A.; Manthrithilake, H.; Daniel, C.; Amann, K.; Zwiener, C.; Barth, J.A.C.; et al. Possible links between groundwater geochemistry and chronic kidney disease of unknown etiology (CKDu): An investigation from the Ginnoruwa region in Sri Lanka. *Expo. Health* **2020**, *12*, 823–834. [CrossRef]

9. Vlahos, P.; Schensul, S.L.; Nanayakkara, N.; Chandrajith, R.; Haider, L.; Anand, S.; Silva, K.T.; Schensul, J.J. Kidney progression project (KiPP): Protocol for a longitudinal cohort study of progression in chronic kidney disease of unknown etiology in Sri Lanka. *Glob. Public Health* **2019**, *14*, 214–226. [CrossRef]

10. Kafle, K.; Balasubramanya, S.; Horbulyk, T. Prevalence of chronic kidney disease in Sri Lanka: A profile of affected districts reliant on groundwater. *Sci. Total Environ.* **2019**, *694*, 133767. [CrossRef]

11. Ludlow, M.; Luxton, G.; Mathew, T. Effects of fluoridation of community water supplies for people with chronic kidney disease. *Nephrol. Dial. Transpl.* **2007**, *22*, 2763–2767. [CrossRef] [PubMed]

12. Wasana, H.M.S.; Aluthpatabendi, D.; Kularatne, W.M.T.D.; Wijekoon, P.; Weerasooriya, R.; Bandara, J. Drinking water quality and chronic kidney disease of unknown etiology (CKDu): Synergic effects of fluoride, cadmium and hardness of water. *Environ. Geochem. Health* **2015**, *38*, 157–168. [CrossRef] [PubMed]

13. Balasooriya, B.M.J.K.; Chaminda, G.G.T.; Weragoda, S.K.; Kankanamge, C.E.; Kawakami, T. Assessment of groundwater quality in Sri Lanka using multivariate statistical techniques. In *Contaminants in Drinking and Wastewater Sources: Challenges and Reigning Technologies*; Kumar, M., Snow, D.D., Honda, R., Mukherjee, S., Eds.; Springer: Singapore, 2021.

14. Edirisinghe, E.A.N.V.; Manthrithilake, H.; Pitawala, H.M.T.G.A.; Dharmagunawardhane, H.A.; Wijayawardane, R.L. Geochemical and isotopic evidences from groundwater and surface water for understanding of natural contamination in chronic kidney disease of unknown etiology (CKDu) endemic zones in Sri Lanka. *Isot. Environ. Health Stud.* **2018**, *54*, 244–261. [CrossRef]

15. Wickramarathna, S.; Balasooriya, S.; Diyabalanage, S.; Chandrajith, R. Tracing environmental aetiological factors of chronic kidney diseases in the dry zone of Sri Lanka—A hydrogeochemical and isotope approach. *J. Trace Elem. Med. Biol.* **2017**, *44*, 298–306. [CrossRef]

16. Nikagolla, C.; Meredith, K.T.; Dawes, L.A.; Banati, R.B.; Millar, G.J. Using water quality and isotope studies to inform research in chronic kidney disease of unknown aetiology endemic areas in Sri Lanka. *Sci. Total Environ.* **2020**, *745*, 140896. [CrossRef]

17. Amarathunga, A.; Kazama, F. Impact of land use on surface water quality: A case study in the Gin river basin, Sri Lanka. *Asian J. Water Environ. Pollut.* **2016**, *13*, 1–13. [CrossRef]

18. Jayasumana, C.; Fonseka, S.; Fernando, P.U.A.I.; Jayalath, K.; Amarasinghe, M.; Siribaddana, S.; Gunatilake, S.; Paranagama, P. Phosphate fertilizer is a main source of arsenic in areas affected with chronic kidney disease of unknown etiology in Sri Lanka. *SpringerPlus* **2015**, *4*, 90. [CrossRef]

19. Chandrajith, R.; Diyabalanage, S.; Dissanayake, C. Geogenic fluoride and arsenic in groundwater of Sri Lanka and its implications to community health. *Groundw. Sustain. Dev.* **2020**, *10*, 100359. [CrossRef]

20. Dharma-Wardana, M.W.C.; Amarasiri, S.L.; Dharmawardene, N.; Panabokke, C.R. Chronic kidney disease of unknown aetiology and ground-water ionicity: Study based on Sri Lanka. *Environ. Geochem. Health* **2015**, *37*, 221–231. [CrossRef]

21. Dissanayake, C.B.; Chandrajith, R. Groundwater fluoride as a geochemical marker in the etiology of chronic kidney disease of unknown origin in Sri Lanka. *Ceylon J. Sci.* **2017**, *46*, 3–12. [CrossRef]

22. Dissanayake, C.B.; Chandrajith, R. Fluoride and hardness in groundwater of tropical regions—Review of recent evidence indicating tissue calcification and calcium phosphate nanoparticle formation in kidney tubules. *Ceylon J. Sci.* **2019**, *48*, 197–207. [CrossRef]

23. Chandrajith, R.; Dissanayake, C.; Ariyarathna, T.; Herath, H.; Padmasiri, J. Dose-dependent Na and Ca in fluoride-rich drinking water—Another major cause of chronic renal failure in tropical arid regions. *Sci. Total Environ.* **2011**, *409*, 671–675. [CrossRef]

24. Xiong, X.-Z.; Liu, J.; He, W.; Xia, T.; He, P.; Chen, X.; Yang, K.; Wang, A. Dose–effect relationship between drinking water fluoride levels and damage to liver and kidney functions in children. *Environ. Res.* **2007**, *103*, 112–116. [CrossRef] [PubMed]

25. Wanigasuriya, K.P.; Peiris-John, R.J.; Wickremasinghe, R. Chronic kidney disease of unknown aetiology in Sri Lanka: Is cadmium a likely cause? *BMC Nephrol.* **2011**, *12*, 32. [CrossRef] [PubMed]

26. Ileperuma, O.; Dharmagunawardhane, H.; Herath, K. Dissolution of aluminium from sub-standard utensils under high fluoride stress: A possible risk factor for chronic renal failure in the North-Central province. *J. Natl. Sci. Found. Sri Lanka* **2009**, *37*, 219–222. [CrossRef]

27. Jayasumana, C.; Gunatilake, S.; Senanayake, P. Glyphosate, hard water and nephrotoxic metals: Are they the culprits behind the epidemic of chronic kidney disease of unknown etiology in Sri Lanka? *Int. J. Environ. Res. Public Health* **2014**, *11*, 2125–2147. [CrossRef]

28. Dissanayake, C.B.; Chandrajith, R. The hydrogeological and geochemical characteristics of groundwater of Sri Lanka. In *Groundwater of South Asia*; Mukherjee, A., Ed.; Springer: Singapore, 2018; pp. 405–428.
29. Hartmann, J.; Moosdorf, N. The new global lithological map database GLiM: A representation of rock properties at the Earth surface. *Geochem. Geophys. Geosyst.* **2012**, *13*, Q12004. [CrossRef]
30. Fick, S.E.; Hijmans, R.J. WorldClim 2: New 1-km spatial resolution climate surfaces for global land areas. *Int. J. Clim.* **2017**, *37*, 4302–4315. [CrossRef]
31. Kawakami, T.; Motoyama, A.; Nagasawa, S.; Weragoda, S.; Chaminda, T. *Groundwater Quality Atlas of Sri Lanka*; Sanduni Offset Printers Kandy: Peradeniya, Sri Lanka, 2014.
32. Robinson, N.; Regetz, J.; Guralnick, R.P. EarthEnv-DEM90: A nearly-global, void-free, multi-scale smoothed, 90m digital elevation model from fused ASTER and SRTM data. *ISPRS J. Photogramm. Remote Sens.* **2014**, *87*, 57–67. [CrossRef]
33. Gong, P.; Liu, H.; Zhang, M.; Li, C.; Wang, J.; Huang, H.; Clinton, N.; Ji, L.; Li, W.; Bai, Y.; et al. Stable classification with limited sample: Transferring a 30-m resolution sample set collected in 2015 to mapping 10-m resolution global land cover in 2017. *Sci. Bull.* **2019**, *64*, 370–373. [CrossRef]
34. WHO. *Guidelines for Drinking-Water Quality*, 4th ed.; World Health Organization: Geneva, Switzerland, 2011; p. 541.
35. Jayathunga, K.; Diyabalanage, S.; Frank, A.; Chandrajith, R.; Barth, J.A.C. Influences of seawater intrusion and anthropogenic activities on shallow coastal aquifers in Sri Lanka: Evidence from hydrogeochemical and stable isotope data. *Environ. Sci. Pollut. Res.* **2020**, *27*, 23002–23014. [CrossRef] [PubMed]
36. Herath, H.M.A.S.; Kubota, K.; Kawakami, T.; Nagasawa, S.; Motoyama, A.; Weragoda, S.K.; Chaminda, G.G.T.; Yatigammana, S.K. Potential risk of drinking water to human health in Sri Lanka. *Environ. Forensics* **2017**, *18*, 241–250. [CrossRef]
37. SLS. *Specification for Bottled (Packaged) Drinking Water*; Sri Lanka Standard 894; Sri Lanka Standard Insitute: Colombo, Sri Lanka, 2020.
38. Bandara, J.; Wijewardena, H.; Liyanege, J.; Upul, M. Chronic renal failure in Sri Lanka caused by elevated dietary cadmium: Trojan horse of the green revolution. *Toxicol. Lett.* **2010**, *198*, 33–39. [CrossRef] [PubMed]
39. Jayatilake, N.; Mendis, S.; Maheepala, P.; Mehta, F.R. Chronic kidney disease of uncertain aetiology: Prevalence and causative factors in a developing country. *BMC Nephrol.* **2013**, *14*, 180. [CrossRef] [PubMed]
40. Nanayakkara, S.; Senevirathna, L.; Harada, K.H.; Chandrajith, R.; Hitomi, T.; Abeysekera, T.; Muso, E.; Watanabe, T.; Koizumi, A. Systematic evaluation of exposure to trace elements and minerals in patients with chronic kidney disease of uncertain etiology (CKDu) in Sri Lanka. *J. Trace Elem. Med. Biol.* **2019**, *54*, 206–213. [CrossRef]
41. Bandara, U.; Diyabalanage, S.; Hanke, C.; van Geldern, R.; Barth, J.A.; Chandrajith, R. Arsenic-rich shallow groundwater in sandy aquifer systems buffered by rising carbonate waters: A geochemical case study from Mannar Island, Sri Lanka. *Sci. Total Environ.* **2018**, *633*, 1352–1359. [CrossRef]
42. Ranasinghe, N.; Kruger, E.; Chandrajith, R.; Tennant, M. The heterogeneous nature of water well fluoride levels in Sri Lanka: An opportunity to mitigate the dental fluorosis. *Community Dent. Oral Epidemiol.* **2019**, *47*, 236–242. [CrossRef]
43. Zeng, J.; Han, G.; Yang, K. Assessment and sources of heavy metals in suspended particulate matter in a tropical catchment, northeast Thailand. *J. Clean. Prod.* **2020**, *265*, 121898. [CrossRef]
44. Amarathunga, U.; Diyabalanage, S.; Bandara, U.; Chandrajith, R. Environmental factors controlling arsenic mobilization from sandy shallow coastal aquifer sediments in the Mannar Island, Sri Lanka. *Appl. Geochem.* **2019**, *100*, 152–159. [CrossRef]
45. Xu, S.; Li, S.-L.; Zhong, J.; Li, C. Spatial scale effects of the variable relationships between landscape pattern and water quality: Example from an agricultural karst river basin, Southwestern China. *Agric. Ecosyst. Environ.* **2020**, *300*, 106999. [CrossRef]
46. Rango, T.; Jeuland, M.; Manthrithilake, H.; McCornick, P. Nephrotoxic contaminants in drinking water and urine, and chronic kidney disease in rural Sri Lanka. *Sci. Total Environ.* **2015**, *518-519*, 574–585. [CrossRef] [PubMed]
47. Wanigasuriya, K. Update on uncertain etiology of chronic kidney disease in Sri Lanka's North-Central dry zone. *MEDICC Rev.* **2014**, *16*, 61–65. [CrossRef] [PubMed]
48. Perera, T.; Ranasinghe, S.; Alles, N.; Waduge, R. Effect of fluoride on major organs with the different time of exposure in rats. *Environ. Health Prev. Med.* **2018**, *23*, 17. [CrossRef] [PubMed]
49. Zhang, J.; Song, J.; Zhang, J.; Chen, X.; Zhou, M.; Cheng, G.; Xie, X. Combined effects of fluoride and cadmium on liver and kidney function in male rats. *Biol. Trace Elem. Res.* **2013**, *155*, 396–402. [CrossRef]

 *water*

MDPI

Article

Effects of Seasonal Thermal Stratification on Nitrogen Transformation and Diffusion at the Sediment-Water Interface in a Deep Canyon Artificial Reservoir of Wujiang River Basin

Yongmei Hou [1,2], Xiaolong Liu [1,*], Sainan Chen [3], Jie Ren [1,2], Li Bai [1], Jun Li [1], Yongbo Gu [1,2] and Lai Wei [1,2]

[1] Tianjin Key Laboratory of Water Resources and Environment, Tianjin Normal University, Tianjin 300387, China; ymhou20@163.com (Y.H.); renjie@tjnu.edu.cn (J.R.); baili@tjnu.edu.cn (L.B.); lijun5931@163.com (J.L.); guyongbo@tjnu.edu.cn (Y.G.); weilai@tjnu.edu.cn (L.W.)
[2] School of Geography and Environmental Science, Tianjin Normal University, Tianjin 300387, China
[3] Institute of Surface—Earth System Science, Tianjin University, Tianjin 300072, China; sainanchen@tju.edu.cn
* Correspondence: Xiaolong.liu@tjnu.edu.cn

Citation: Hou, Y.; Liu, X.; Chen, S.; Ren, J.; Bai, L.; Li, J.; Gu, Y.; Wei, L. Effects of Seasonal Thermal Stratification on Nitrogen Transformation and Diffusion at the Sediment-Water Interface in a Deep Canyon Artificial Reservoir of Wujiang River Basin. *Water* 2021, 13, 3194. https://doi.org/10.3390/w13223194

Academic Editor: Bommanna Krishnappan

Received: 20 September 2021
Accepted: 9 November 2021
Published: 11 November 2021

Publisher's Note: MDPI stays neutral with regard to jurisdictional claims in published maps and institutional affiliations.

Abstract: Watershed-scale nitrogen pollution in aquatic systems has become a worldwide concern due to its continuous impact on water quality deterioration, while the knowledge of key influencing factors dominating nitrogen transportation and transformation at the sediment-water interface (SWI) remains limited, especially in impounded rivers with an artificial reservoir. Hence, for a better understanding of the effects of thermal stratification on nitrogen transformation, we investigated the nitrogen species and isotopes in the sediment of a deep reservoir in Southwest China. Our results confirmed a significant difference in nitrogen species and isotopic composition in sediment between those in the thermal stratification period and non-thermal stratification period and indicated that the sediment biogeochemical process and transportation were clearly linked to the variations in water temperature and dissolved oxygen dominated by the process of thermal stratification. Significant seasonal differences in NH_4^+-N and NO_3^--N in pore water of the upper layer (0–19 cm) revealed that nitrification exhausted NH_4^+ in the non-stratified period (NSP), and a potential low mineralization rate appeared when compared with those in the stratified period (SP). Seasonal differences in nitrogen species and isotope fractionation of δ^{15}N-PON (about 2.3‰ in SP) in the upper layer sediment indicated a higher anaerobic mineralization rate of organic matter in SP than that in NSP. The diffusion fluxes of NH_4^+-N at SWI were 9.48 and 15.66 mg·m^{-2}·d^{-1} in NSP and SP, respectively, and annual NH_4^+-N diffusion accounted for 21.8% of total storage in the reservoir. This study demonstrated that the nitrogen cycling processes, especially nitrification, denitrification, and mineralization, have been largely altered along with the changes in dissolved oxygen and that the diffusion of nitrogen species varied with the presence of the oxygen. The results contribute to the future study of watershed nitrogen budget evaluation and suggest that the endogenous nitrogen released from the sediment-water interface should be emphasized when aiming to fulfil water management policies in deep reservoirs.

Keywords: thermal stratification; nitrogen transformation; sediment-water interface; reservoir; NH_4^+ diffusion

1. Introduction

Excessive nitrogen loading at watershed-scale in aquatic systems, with the constant input of anthropogenically sourced active nitrogen, has posed serious environmental problems, such as eutrophication, greenhouse gas emission, and water quality deterioration [1,2]. Watershed-scale nitrogen pollution persistently attracts worldwide concern attributed to the enormous challenge of nitrogen assessment and management in global ecosystems. By intercepting runoff and sediment, massive dams and reservoirs worldwide would deposit the dissolved and adsorbed nutrients with the sediment and conserved within the reservoir, which tends to reduce levels of nutrient transportation to downstream rivers and instead accumulate the nutrients in the reservoir [3,4]. Moreover, recent studies have emphasized that the accumulation of nitrogen in aquatic systems was affected by external nitrogen input as well as internal nitrogen release, which is dominated by nitrogen biogeochemical processes [5,6]. Evaluating the transformation and transportation of nitrogen species among the sediment-water interface (SWI) is essential if we aim to reduce nitrogen pollution in a reservoir and help to establish relevant policies for reservoir water quality protection. However, few studies have assessed the contribution of nitrogen releasing from the SWI in the reservoir. Currently, an increasing number of studies have raised the controversies surrounding the role of sediment in nitrogen transportation and transformation at the SWI, as well as its influence on the fate and distribution of nitrogen afterwards, which has become a hot topic in academic community [7,8].

Nitrogen experiences complex biogeochemical processes under different environmental conditions, e.g., DO has been recognized as the crucial factor that dominates nitrification and denitrification, and dissimilatory nitrate reduction to anammox (DNRA) generally occurs in sediment with a high C/N ratio and C loading [9]. Generally, ammonification, denitrification, and DNRA were regarded as being the dominant processes in a hypoxic environment in the stratified period (SP), while ammonification and nitrification could be favored by oxygenated conditions in the non-stratified period (NSP). In recent decades, intensive demands for hydropower resources led to the emergence of special landscape features caused by the cascading development in most rivers [10,11], which significantly posed a challenge for the synchronously high-efficiency management of water resources and the water environment. Over the past 60 years, approximately 58,000 large reservoirs have been built worldwide [12], which altered the original ecosystem conditions and attracted widespread concern [13].

It was well known that there was seasonal thermal stratification in deep canyon reservoirs, and water temperatures in the hypolimnion of deep lakes/reservoirs were generally low (4–6 °C) and remained nearly constant [13,14]. However, temperature-driven hypoxia in the stratified period (SP) and oxygenated hypolimnion in the non-stratified period (NSP) may dominate the nitrogen exchange between sediment and water. Muller et al. [15] indicated that, in a lake, anaerobic environmental conditions were more conducive to internal nitrogen release than those in aerobic conditions. In addition, some studies reported that T and DO were the key factors controlling endogenous nitrogen diffusion of sediment in a shallow lake and that the diffusion fluxes of NH_4^+ were positively correlated with T, while there was a negative correlation with DO [16]. Cai et al. [17] investigated the relationship between nitrogen species and biological activities in shadow lake sediment, and reported that T and DO significantly alter the abundance of bacterial genes and community structures, subsequently affecting the transformation processes of sediment nitrogen. Hence, in this study, we hypothesized that the seasonal variation in DO and T would dominate the nitrogen transformation at SWI in a deep reservoir. However, as noted above, most current studies focused on nitrogen diffusion and influencing factors in thermal stratified lakes, while few studies have conducted targeted research in deep canyon reservoirs, especially the deep reservoirs in Southwest China. Because an artificial reservoir has a special hydrological regulation and water management regime for the purpose of power generation, the particulate matter sources and accumulation rates into the sediment are different from those in lakes. In order to optimize water protection and reduce nitrogen accumulation in reservoirs, sediment nitrogen transformation and diffusion to overlying water—which tends to play a role in endogenous pollution—urgently needed to be understood.

Water-soluble inorganic nitrogen and adsorbed inorganic nitrogen were the dominant nitrogen forms at the SWI; a growing number of studies have documented that the water-soluble and adsorbed nitrogen varied significantly along with the water chemistry changes [18]. Generally, there are two methods that have been employed universally to study nitrogen cycling at the SWI: incubation of intact sediment core experiments and profiling of pore waters coupled to deterministic models [19,20]. The former methodology has revealed that the nitrogen species fluxes at the SWI varied among different aquatic environments, but it is difficult to infer the detailed relevant transformation processes involved at different depths of sediment. For the purpose of both identifying the nitrogen cycling processes and estimating the effluxes and influxes of nitrogen species at the SWI, the later methodology could provide more information on the nitrogen fate at the SWI. Additionally, stable isotopes approaches have been applied extensively in recent studies on watershed-scale nitrogen cycling for better understanding of the nitrogen transformation processes [21–23] and for identifying the source of particulate matter.

Most current studies on nitrogen transformation and diffusion of sediment paid close attention to aquatic systems, such as rivers, lakes, and oceans, while the knowledge on the fate of nitrogen and related key influencing factors at the sediment and water interface in deep reservoirs is still limited. In order to understand the effects of seasonal thermal stratification on nitrogen transformation and to quantitatively evaluate the contamination of endogenous nitrogen in a deep reservoir, the concentrations of nitrogen species and stable isotopes (δ^{15}N-PON) of particulate organic matter were analyzed in sediments, which were collected during SP and NSP in Wujiangdu (WJD) reservoir, a typical deep canyon eutrophic reservoir in Southwest China. The study aimed to (1) clarify the seasonal variation characteristics of sediment nitrogen species in a deep reservoir; (2) to understand how the DO affected the transformation and diffusion of sediment nitrogen by comparative study between SP and NSP; (3) to quantitatively estimate the nitrogen exchanges at the SWI and assess the influence of seasonal thermal stratification. The results of this study provide implications useful for assessing the effects of endogenous nitrogen on nitrogen contamination in a deep reservoir that has characteristics of seasonal thermal stratification and nitrogen transformation and transportation that are similar to other deep reservoirs worldwide, and helping to fulfill water management policies in watersheds in consideration of the nitrogen contamination originating from the SWI.

2. Materials and Methods

2.1. Study Area

Wujiang River Basin is one of the largest tributaries on the south bank of the Yangtze River, and the WJD reservoir is the earliest (1979) seasonal regulating hydropower station built and put into use in China. The WJD reservoir is located in the middle reaches of the Wujiang River Basin (106°47′ E, 27°18′ N, 27,790 km^2), which drains the Southeast Asian Karst Region [24,25]. The average annual flow rate of the WJD reservoir is 502 m$^3 \cdot$s^{-1}, and the storage area is 47.8 km^2. There are abundant coal resources and industrial production bases in the area, such as a Wujiang manganese steel plant, phosphate fertilizer plant, and chemical fertilizer plant. Historically, intensive human activity has caused a serious eutrophication in the WJD reservoir, mainly on account of perennial cage fish aquaculture; hence, the hydrochemistry of the WJD reservoir is characterized by its high nitrogen loading [5].

The WJD reservoir, a deep canyon artificial reservoir, had significant seasonal variation in the profiles of water chemical parameters (especially the water T and DO). The WJD reservoir becomes thermally non-stratified from November to March, while remaining thermally stratified from April to October [5]. The water T difference between upper and lower layers reaches to 12.4 °C during the SP, while the water profile is homogeneous in the NSP, with small temperature differences (0.11 °C) among the vertical water mass. In Spring (typically in April) and Autumn (in October), the water layers are in a transitional stage of thermal stratification, and the most significant thermally stratification and non-stratification occurs in July and January, respectively. Thus, it was ideal to study the influences of seasonal thermal stratification on nitrogen transformation and transportation in the sediment-water interface by comparing those in July and January. The sediment type in the WJD reservoir is fine-grained, with a high proportion of clay [26]. The water chemical parameters in the WJD reservoir are shown in Table 1.

Table 1. Water chemical parameters in the WJD reservoir from our previous study [27].

	T (°C)	DO (mg·L^{-1})	EC (ms·cm^{-1})	pH	NH4$^+$-N (mg·L^{-1})
Range	10.7~28.6	6.6~15.9	0.18~0.32	7.8~8.9	0.01~1.44
	K$^+$ (mg·L^{-1})	Na$^+$ (mg·L^{-1})	Ca^{2+} (mg·L^{-1})	Mg^{2+} (mg·L^{-1})	NO$_3$$^-$-N (mg·L^{-1})
Range	1.56~1.81	3.36~4.38	61.32~65.64	10.78~11.80	1.23~3.50

2.2. Sampling and Analysis

A sampling site was chosen at the front of the dam (<1 km to the dam) in Wujiangdu Reservoir, where the depth of the water reached to 113 m and 95 m in the NSP and SP, respectively. Based on previous studies that indicated that sediment cores in lakes and reservoirs are generally consistent, single or two sediment cores would thus be enough for studying the transformation and spatial distribution of nutrients and heavy metals in a lake or reservoir [28–31]. Moreover, this study focused on nitrogen transformation and transportation mainly in the deepest parts of the reservoir (close to the dam), where the area was relatively small and the sediment was consistent. Referring to previous studies [28–31], two sampling campaigns were conducted on the NSP (January 2018) and SP (July 2019), respectively, and two sediment cores were collected by a sediment corer (Hydro-Bios, Inc., Altenholz, Germany) at the front of dam in the reservoir (<1 km from the dam), which is the deepest site in the reservoir (Figure 1). The sediment core collected in the NSP was 58 cm long and that in SP was 52 cm; the diameter of the sampling tube was 7 cm. Sample cores were sliced and sub-sampled immediately after sampling in the field (the first 20 cm was sliced at a 1 cm depth interval and at 2 cm intervals afterwards), then stored in anaerobic sample tubes, as in the studies of Hogarh et al. and Copetti et al. [28,31]. These samples were centrifuged by a centrifugal machine placed in a N$_2$-filled container to separate pore water and were then filtered with a 0.45 μm cellulose acetate membrane. After these pretreatment process, they were stored at 4 °C away from light. In the laboratory, after the centrifuged sediment samples were freeze-dried at −30 °C, they were ground with an agate mortar, passed through a 100-mesh sieve, and then stored in polyethylene bags under 4 °C till analyzed. The overlying waters were sampled by two methods. First, for the overlying water close to the sediment (depth < 50 cm), we collected the water samples using a silicone tube (diameter 6 inches) by siphon method in the sediment corer before sediment subsampling; second, the overlying water samples above the sediment (>1 m) were sampled with a water sampler (Hydro-Bios, Altenholz, Germany).

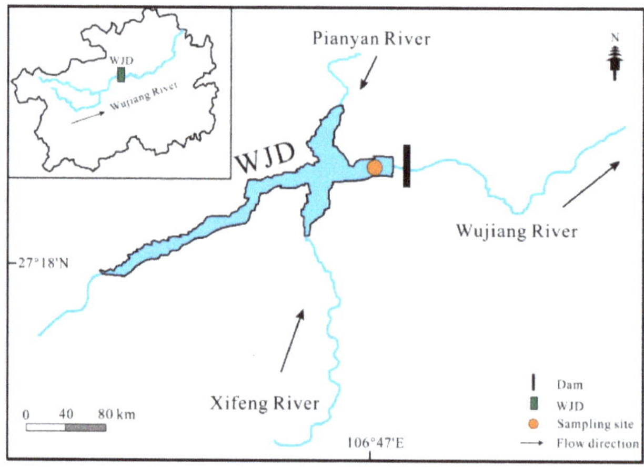

Figure 1. Map showing the study area and the sampling sites at the WJD reservoir.

Water-soluble nitrogen was extracted from 2 g of the ground sediment samples using 10 mL deionized water, then 10 mL 2 mol·L^{-1} KCl was added and processed with the same method to obtain the absorbed inorganic nitrogen—the potential nitrogen of which could be replaced by special iron. Then, the concentrations of nitrogen species (NH_4^+-N, NO_3^--N, and total nitrogen, TN) of pore waters and water-soluble and absorbed nitrogen species were determined by continuous-flow analysis (AA3 Auto Analyzer, SEAL, Norderstedt, Germany). The nitrogen detection limit was 0.001, 0.003, 0.003, and 0.01 mg·L^{-1} for NO_2^--N, NO_3^--N, NH_4^+-N, and TN, respectively. NO_2^--N was under the detection limit. Laboratory standards and replicated samples were employed to keep the precision of NH_4^+, NO_3^-, and TN concentration analysis higher than ±5%. The dissolved organic nitrogen (DON) had a differential value to TN and total inorganic nitrogen (TIN, including NH_4^+-N and NO_3^--N).

Processed sediment samples were weighed at 0.5 g and 15 mL 0.5 mol·L^{-1} HCl and 2 mol·L^{-1} KCl solution was added to remove inorganic carbon and nitrogen, then samples were washed repeatedly with deionized water until becoming neutral. The samples were freeze-dried and ground afterwards for nitrogen-stable isotopes (δ^{15}N-PON) and C/N molar ratio analysis by stable isotopic mass spectrometer (MAT-253) and elemental analyzer (Elementar, Rhine main, Germany) at Tianjin Normal University.

2.3. Calculation of Nitrogen Diffusion at SWI

The diffusion of water-soluble nitrogen at the SWI was driven by the nitrogen gradients and calculated by Fick's first law, following equation [19,20]

$$F = D_S \times \varphi \times (\partial c/\partial z) \qquad (1)$$

where F (mg·m^{-2}·d^{-1}) is the diffusion flux of nitrogen, and $\partial c/\partial z$ (mg·L^{-1}·cm^{-1}) is the gradient of dissolved substance across the SWI, which could be fitted by exponential function using the nutrient concentration of overlying water and the depth of SWI of about 5 cm. Then, the derivative of exponential function when Z = 0 was taken as the concentration gradient. C (mg·L^{-1}) was the concentration of nitrogen, and Z (cm) was the vertical distance, which started from the upper boundary of the SWI and increased with depth. φ is the surface sediment porosity and can be approximately estimated by following equation [32]:

$$\varphi = (r - s)/r \qquad (2)$$

where r is the wet weight of sediment, while s is the dry weight. D_S (cm^2·s^{-1}) is the actual diffusion coefficient under different T conditions, which can be expressed as follows [32,33]:

$$D_S = \varphi\, D_0 \;(\varphi < 0.7) \qquad (3)$$

$$D_S = \varphi^2\, D_0 \;(\varphi \geq 0.7) \qquad (4)$$

where D_0 is the diffusion coefficient in infinite dilution, and the values for NO_3^--N and NH_4^+-N at 25 °C are 19.0×10^{-6} and 19.8×10^{-6} cm^2·s^{-1}, respectively. Thus, it could be corrected according to the following formula due to the different T [34]:

$$D_0 \,(NO_3^-\text{-N}) = 19.0 \times 10^{-6} + 0.4 \times (T - 25\,^\circ C) \qquad (5)$$

$$D_0 \,(NH_4^+\text{-N}) = 19.8 \times 10^{-6} + 0.4 \times (T - 25\,^\circ C) \qquad (6)$$

2.4. Data Processing

IBM SPSS Statistics 19 was used for data analysis. Analysis of variance (ANOVA) was performed to examine the differences in the nitrogen concentrations in the SP and NSP. Additionally, *t*-tests were used to identify the variables of the C/N ratio and δ^{15}N-PON in sediments of before and after artificial dam construction. Significance levels were reported to be $p < 0.05$ and $p < 0.001$. Microsoft Excel was used to analyze the linear correlation of variables and obtain the value of the correlation coefficient (R^2). All the data were completed by Origin 2018, Grapher 15, and Microsoft Office 2010. The concentrations of nitrogen species in the sediment of the WJD reservoir are listed in Table 2.

Table 2. The concentrations of nitrogen species in sediment of WJD reservoir.

Sampling Period	Types	Nitrogen Species	Min (mg·kg^{-1})	Max (mg·kg^{-1})	Average (mg·kg^{-1})	SD	Coefficient of Variation
SP	water-soluble nitrogen	NH_4^+-N	12.15	31.37	21.01	5.12	0.24
		TN	63.51	121.36	86.83	14.34	0.17
	absorbed nitrogen	NH_4^+-N	58.07	85.71	70.99	7.65	0.11
		NO_3^--N	4.81	12.70	7.16	2.28	0.32
		TN	90.95	255.65	159.20	45.35	0.28
NSP	water-soluble nitrogen	NH_4^+-N	5.74	27.64	14.86	4.44	0.30
		NO_3^--N	0.06	0.58	0.31	0.14	0.46
		TN	57.39	339.36	144.49	62.43	0.43
	absorbed nitrogen	NH_4^+-N	48.04	89.97	71.93	9.75	0.14
		NO_3^--N	5.78	24.75	12.67	4.75	0.37
		TN	120.68	398.65	223.85	63.77	0.28

3. Results

3.1. Seasonal Variation of Inorganic Nitrogen in Sediment Pore Water

The concentrations of NH_4^+-N in pore water ranged from 5.7 to 27.6 mg·L^{-1} and from 15.3 to 31.4 mg·L^{-1} (averaged 14.9 and 21.0 mg·L^{-1}) in the NSP and SP (Table 2), respectively. There were significant seasonal differences in vertical distribution of NH_4^+ and NO_3^- in the upper layer (0–19 cm) of the sediment between the two periods, but the distribution nearly synchronously varied at deep layers (>19 cm) (Figure 2). The concentrations of NH_4^+-N were lower in the upper layer of the sediment, then increased with depth in the NSP, while reversing in the SP. Inorganic nitrogen in pore water was mainly composed of NH_4^+-N, accounting for 90.8~99.7% in the NSP. On the contrary, the concentration of NO_3^--N in sediment pore water was only determined at 0–19 cm depth (the concentration ranged from 0.056 to 0.58 mg·L^{-1}) in the NSP, while it was under the detection limit (0.001 mg·L^{-1}) in the SP and the deep layer of the sediment in the NSP. NO_3^--N was detected only in the NSP at the depth of 0–19 cm with a low concentration in pore water, while the concentration of NH_4^+-N notably exceeded NO_3^--N ($p < 0.001$) in this period.

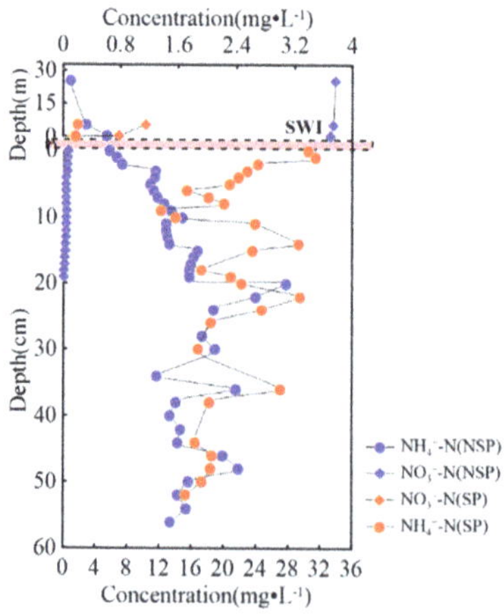

Figure 2. Characteristics of the vertical distribution of inorganic nitrogen in sediment pore water.

3.2. Seasonal Changes in Water-Soluble and Adsorbed Nitrogen in Sediment

Concentrations of adsorbed total nitrogen (TN) ranged from 120.7 to 398.56 mg·kg^{-1} in the NSP and from 91 to 255.7 mg·kg^{-1} in the SP, with an average value of 233.9 and 159.2 mg·kg^{-1}, respectively. The water-soluble TN ranged from 57.4 to 339.4 mg·kg^{-1} and from 63.5 to 121.36 mg·kg^{-1} (averaged 144.5 and 86.8 mg·kg^{-1}) in the NSP and SP (Table 2), respectively. The adsorbed TN was 1.4 to 2.7 times higher than that of the water-soluble TN. Meanwhile, the concentrations of nitrogen species displayed different seasonal variation; adsorbed TN in the sediment at 0–10 cm in the SP increased with depth, and there was no significant change in water-soluble TN (Figure 3).

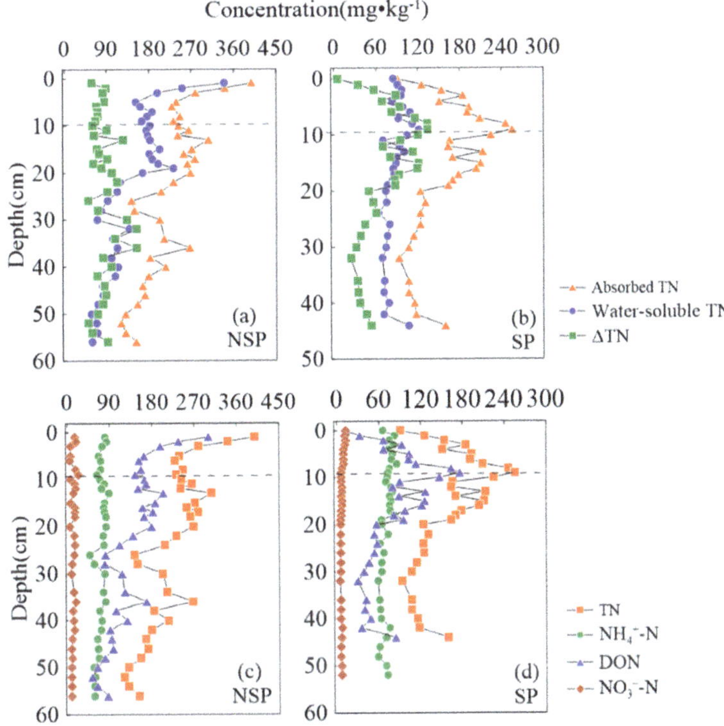

Figure 3. Characteristics of the vertical distribution of absorbed nitrogen species in sediment.

The vertical distribution of adsorbed TN in two seasons was roughly the same as those of DON, and there were obvious seasonal differences in the upper layer (0–10 cm) of sediment (Figure 3). In addition, DON concentrations were higher and decreased with depth in the NSP, while displaying the opposite in the SP (Figure 4). Figure 4 shows the vertical variation of different nitrogen species in both the SP and NSP, with the statistics of each nitrogen species including the whole data of the profile. Moreover, the concentration of adsorbed inorganic nitrogen varied in a smaller range than that in DON, and NH$_4^+$-N was the main component, accounting for 80.5% to 93.3% of the TN. The concentrations of adsorbed NH$_4^+$-N ranged from 48.0 to 90.1 mg·kg^{-1} in the NSP and from 58.1 to 85.7 mg·kg^{-1} in the SP, while the concentrations of absorbed NO$_3^-$-N ranged from 5.8 to 24.75 mg·kg^{-1} in the NSP and from 4.8 to 12.7 mg·kg^{-1} in the SP. DON was the dominant component of TN (Figure 4), ranging from 52.8 to 300.1 mg·kg^{-1} in the NSP and from 13.0 to 176.4 mg·kg^{-1} in the SP, with a mean proportion ranging from 47% to 60%.

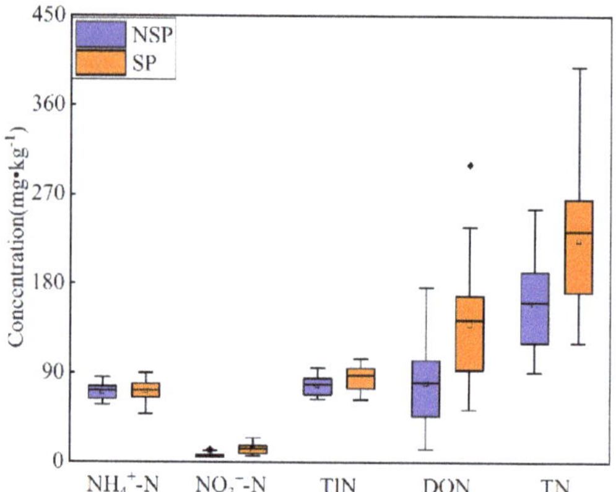

Figure 4. Concentrations of absorbed nitrogen species in the WJD reservoir.

3.3. Characteristics of the Vertical Distribution of Stable Nitrogen Isotope (δ^{15}N-PON) and the C/N Ratio in Sediment Particulate Organic Matter

The profiles of the C/N ratio displayed no obvious seasonal variation when increasing with depth (Figure 5), and the values varied from 9.1 to 16.2 (mean value was 11.8) in the NSP and from 10.1 to 15.3 (mean value was 12.2) in the SP. The δ^{15}N-PON values ranged from +5.7‰ to +6.8‰ and from +4.9‰ to +7.2‰, with an average value of +6.3‰ and +5.7‰ in the NSP and SP, respectively. Additionally, the δ^{15}N-PON values decreased slightly when compared with those from 10 years ago (from +6.84‰ to +13.64‰), and there was inconspicuous difference in the vertical distribution of δ^{15}N-PON in the NSP. Contrarily, it was roughly divided into two stages in the SP: the first stage was 0–15 cm (the mean value was +6.0‰, with the highest value of +7.2‰ and the lowest value of +4.9‰) and generally decreased with depth; the second stage was 16–52 cm (mean value was +5.4‰), where the δ^{15}N-PON varied in a small range.

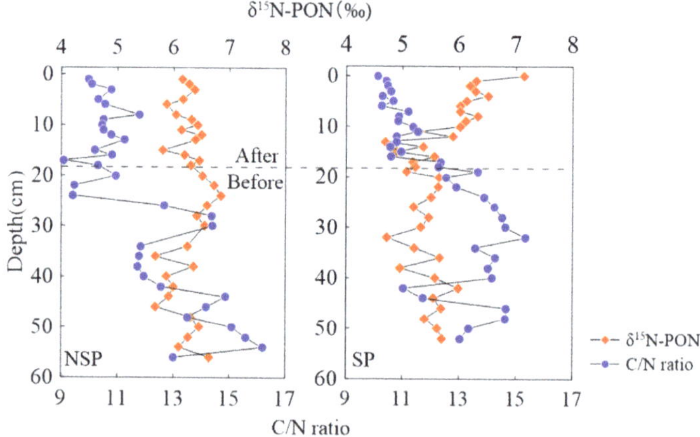

Figure 5. Characteristics of the vertical distribution of δ^{15}N-PON and C/N ratio in particulate organic matter.

3.4. Diffusion Fluxes of Inorganic Nitrogen at the SWI

Considering the seasonal differences in DO and T conditions in a deep reservoir, the actual T at SWI during the sampling period (14.7 °C in the NSP and 12.2 °C in the SP) was used to correct the actual diffusion coefficients of NH_4^+-N and NO_3^--N (Table 3). The diffusion fluxes (F) of inorganic nitrogen during the sampling period were calculated with $\partial c/\partial z$ and φ (Table 4), and the values for NH_4^+-N were 9.48 and 15.66 mg·m^{-2}·d^{-1}; for NO_3^--N the values were -1.49 and -0.21 mg·m^{-2}·d^{-1} in the NSP and SP, respectively. Generally, a positive flux represents nitrogen released from sediment to overlying water, while a negative flux represents that from the overlying water to the sediment.

Table 3. The actual diffusion coefficient under different T conditions of NH_4^+-N and NO_3^--N.

T (°C)	NH_4^+-N (cm^2·s^{-1})	NO_3^--N (cm^2·s^{-1})
0	9.8×10^{-6}	9.78×10^{-6}
12.2	9.87×10^{-6}	9.33×10^{-6}
14.7	10.54×10^{-6}	10.01×10^{-6}
25	19.8×10^{-6}	19×10^{-6}

Table 4. Calculation of inorganic nitrogen diffusion fluxes at the SWI of WJD.

Sampling Time	Inorganic Nitrogen	Exponential Function	R^2	φ	$\partial c/\partial z$ (mg·L^{-1}·cm^{-1})	F (mg·m^{-2}·d^{-1})
January 2018	NH_4^+	$y = 3.2791e^{-0.368x}$	0.88	0.82	-1.27	9.48
	NO_3^-	$y = 0.838e^{0.2492x}$	0.92		0.21	-1.49
July 2019	NH_4^+	$y = 6.3376e^{-0.353x}$	0.73		-2.24	15.66
	NO_3^-	$y = 0.24e^{0.1345x}$	0.63		0.032	-0.21

4. Discussion

4.1. Effects of Seasonal Thermal Stratification on Nitrogen Transformation in Sediment Pore Water

In the sediment of the WJD reservoir, the variations in the concentrations of NO_3^--N and NH_4^+-N in pore water suggested that the proportion of NH_4^+ consumed by sediment nitrification was much lower than the NH_4^+ generated by mineralization under hypoxic conditions. Additionally, the significantly higher concentration of NH_4^+-N in pore water during the SP than that of the NSP ($p < 0.05$) indicated that the NH_4^+ was mainly controlled by anaerobic mineralization. The seasonal differences in concentrations of NH_4^+-N and NO_3^--N were mainly driven by DO level, since the oxygenated hypolimnetic water would promote the NH_4^+ that originated from mineralization to be nitrified to NO_3^- [5,14]. Moreover, the seasonal differences in the vertical distribution of NH_4^+ and NO_3^- in the upper layer (0–19 cm) and deep layers (>19 cm) (Figure 2) indicated that the effects of seasonal thermal stratification on NH_4^+-N in pore water only played a role in the upper layer (Figure 6). This might be related to the sediment porosity, which only permitted DO to permeate to some extent.

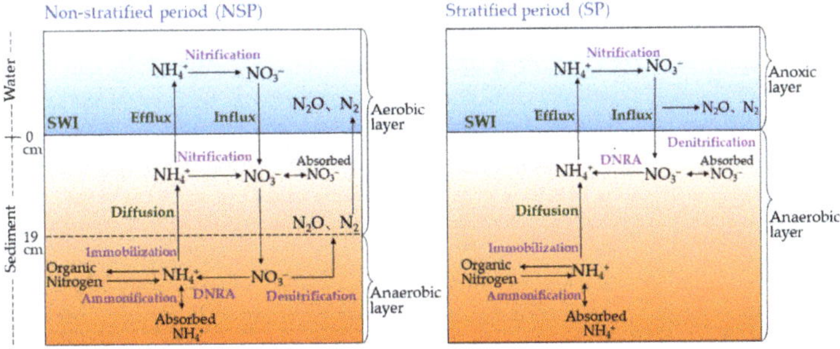

Figure 6. Schematic diagrams of major nitrogen biogeochemical processes at the SWI during the NSP and SP at the WJD reservoir.

Previous studies have demonstrated that the concentrations of NH_4^+-N in the pore water of sediment are dominated by the content of organic matter, degradation rate, and deposition environment [6]. What is more, it has been verified that anaerobic mineralization is mainly affected by sediment properties (especially porosity) [35]. Additionally, a number of studies revealed that larger porosity in upper layers of the sediment would provide better conditions for various microorganism-driven nitrogen processes than that in deep layers, mainly resulting from better oxygen and nitrogen exchange among the SWI. Thus, it could be inferred that the concentration of NH_4^+-N decreased with depth in the upper layer in the SP, mainly owing to the different porosity between upper layers and deep layers by affecting the potential ammonification of organic matter in the two layers.

Subsequently, it has been reported that the dissimilatory nitrate reduction to ammonium (DNRA) could easily occur under the proper conditions with high carbon content and limited NO_3^- in the sediment [9,36], which is consistent with the findings of this study in the WJD reservoir (carbon content reached 5.05%, and the concentrations of NO_3^--N were less than 0.31 mg·L^{-1}). Additionally, along with the NO_3^- decreasing in the upper layer in the NSP, the NH_4^+ increased dramatically under hypoxic-anaerobic conditions (Figure 2), and the aerobic mineralization should not contribute so much newly produced NH_4^+. Thus, there might be a proportion of newly produced NH_4^+ originating from DNAR. Overall, the shift of thermal stratification from the SP to the NSP not only contributed NO_3^- in the SWI but also changed the nitrogen cycling processes of the sediment upper layer, i.e., the nitrogen transformation was jointly influenced by nitrification, mineralization, and DNAR in the NSP (Figure 6). Based on the discussion and findings mentioned above, we inferred the major nitrogen biogeochemical processes in the WJD reservoir by using schematic diagrams, as shown in Figure 6. In this study, different nitrogen biogeochemical processes were addressed in the SP and NSP, respectively. With that, we want to highlight the contribution of nitrification to the release of nitrogen at the SWI in the NSP since it will consequently influence nitrogen transportation in the downstream rivers, which will probably lead to uncertainty when evaluating the nitrogen budget in an impounded river.

4.2. Contribution of Absorbed Nitrogen Species in Sediment Nitrogen Cycling

Concentrations of adsorbed TN were significantly higher than those of water-soluble TN ($p < 0.05$) in the SP and NSP, which was mainly attributed to the clay minerals in the sediment adsorbing a large amount of ionic nitrogen. Here, we employed the differences between water-soluble and absorbed TN (ΔTN) to characterize the potential contribution and ability of absorbed nitrogen species in particulate matter to directly participate in the nitrogen transformation process (Figure 3). A clear seasonal variation in ΔTN was observed (ANOVA, N = 10, $p = 0.001$), especially in the upper layer (0–10 cm) of sediment. Compared with the constant variation in ΔTN during the NSP, the increasing ΔTN in the upper layer during the SP (Figure 3) suggested that it was more conducive for adsorbed nitrogen to participate in the conversion of water-soluble nitrogen in the SP. Furthermore, in the deep layers (>10 cm), the similar decreases and comparable concentrations of ΔTN, water soluble TN, and adsorbed TN between the two periods indicated that the adsorbed nitrogen contributed less to soluble TN than that in the upper layer and that the effects of thermal stratification on the transformation of adsorbed nitrogen was limited.

Specifically, the increases in ΔTN and adsorbed TN in the upper layer during SP mainly resulted from the increases in dissolved organic nitrogen (DON), which was the major component of TN, displaying similar vertical profiles and seasonal variation as adsorbed TN (Figure 3). As a result, the significant seasonal differences in DON could account for those in adsorbed TN (Figure 4), which verified that the seasonal changes in adsorbed TN were primarily controlled by DON. Moreover, the water regulation of the reservoir would affect organic matter in sediment, e.g., anti-seasonal hydrological regime made it possible to continuously accumulate organic particulate matter from upstream to the sediment upper layer in the NSP, while the water discharge in the SP decreased the rate of particulate matter sedimentation, which resulted in an overall higher DON concentration in the NSP than that in the SP ($p < 0.05$). In addition, higher concentrations of NH_4^+ in the sediment upper layer (0–10 cm) in pore water in the SP (Figure 2) and lower concentrations of absorbed DON than those in the NSP (Figure 3) were observed, which were driven by intensive mineralization in the sediment upper layer in the SP, demonstrating that the seasonal differences in mineralization rate were the dominant factor of DON. Furthermore, steady decreases in adsorbed DON in the upper layer (0–10 cm) (Figure 3), accompanied by an increase in NH_4^+ in pore water in the SP (Figure 2) was mainly attributed to the vertical variation in the mineralization rate that generally would be decided by the porosity of the sediment [35]. Additionally, the inconspicuous differences in vertical distributions of absorbed inorganic nitrogen (NH_4^+, NO_3^-) during sampling time (Figure 3), compared with the significant seasonal variations in NH_4^+ in sediment pore water, especially in sediment upper layer (Figure 2), demonstrate that these parts of nitrogen could not participate directly in the nitrogen transformation process. As noted above, the findings of this study revealed the complexity of the nitrogen species distribution and transformation in a thermally stratified reservoir, which is obviously different from natural lakes, and is something that should be emphasized in the future studies of the nitrogen budget and nitrogen management in reservoirs.

4.3. Impact of Dam Construction on Sedimentary Nitrogen Sources

The source of sediment organic matter and nitrogen transformation processes could be effectively characterized by the combination of δ^{15}N-PON and the C/N ratio [37,38]. According to the annual deposition rate (0.46 cm·a^{-1}) [39] and the age of the WJD reservoir, we inferred that sedimentation after dam construction concentrated at the depth of 0–19 cm. The significant differences in sedimentation and nitrogen transformation processes of particulate matter before and after artificial dam construction are displayed by the C/N ratio ($p < 0.001$) and δ^{15}N-PON ($p < 0.05$) (Table 5), respectively. These remarkable differences in the C/N ratio and δ^{15}N-PON imply that dam construction severely alters particulate matter deposition in a reservoir, suggesting that it would be important to reassess the effects of dams and reservoirs on nitrogen accumulation and transportation in impounded rivers.

Table 5. Statistics of C/N ratio and δ^{15}N-PON at different sampling periods.

Sampling Time	Items	Before Construction	After Construction	*t*-Test
January 2018	C/N ratio	11.11 ± 1.68	8.99 ± 0.51	5.09 ***
	δ^{15}N-PON	6.29 ± 0.35	6.22 ± 0.21	0.62
July 2019	C/N ratio	11.72 ± 0.97	9.45 ± 0.74	8.09 ***
	δ^{15}N-PON	5.46 ± 0.31	5.81 ± 0.62	−2.23 *

Note: *** was $p < 0.001$; * was $p < 0.05$.

A large isotope fractionation (+4.9~+7.2‰) was observed at 0–19 cm in the SP. Generally, the different isotopic fractionation among nitrogen cycling processes would verify the values of isotopic signals [21], e.g., the isotope fractionation caused by degradation was about 1‰. Hence, this large isotope fractionation in the SP could be attributed to the mineralization rate under anaerobic conditions. Additionally, the vertical profiles of δ^{15}N-PON displayed a significant seasonal change at 0–19 cm, and it decreased with depth in the SP, while not expressing clear variation in the NSP (Figure 5). Nitrogen isotope fractionation during mineralization leading to ^{15}N enriched in the residual of mineralization (PON) would consequently result in the enrichment of ^{15}N in organic particulate matter at the sediment upper layer. Furthermore, accompanied by an incompatible exponential decay model of depth and the content of organic nitrogen ($R^2 < 0.5$) [35], it was commonly verified that the nitrogen transformation processes were mainly dominated by mineralization at the sediment upper layer under anaerobic conditions.

Lower C/N ratio values were noticed at the sediment upper layer than at the bottom (Figure 5), which could be attributed to DO and microbial activity playing key roles at the sediment upper layer as the major factors for the degradation process of organic matter [40]. Moreover, the vertical distribution of the C/N ratio was displayed analogically in the SP and NSP (Figure 5). On account of the C and N in the degradation processes displaying an equal proportion, which was related to Formulas (7) and (8) [41], the degradation reactions could be considered as a potential factor driving the vertical distribution of the C/N ratio.

$$(CH_2O)_{106}(NH_3)_{16}(H_3PO_4) + 138O_2 \rightarrow 106CO_2 + 16HNO_3 + H_3PO_4 + 122H_2O \quad (7)$$

$$(CH_2O)_{106}(NH_3)_{16}(H_3PO_4) + 94.4HNO_3 \rightarrow 106CO_2 + 55.2N_2 + H_3PO_4 + 177.2H_2O \quad (8)$$

4.4. Endogenous Release of Sediment Nitrogen

Diffusive fluxes of inorganic nitrogen were different depending on seasonal changes, i.e., the diffusion flux (efflux) of NH_4^+-N was 1.65 times in the SP than that in the NSP (Table 4), which was mainly attributed to the concentration gradient and microbial activity verified by the finding that the release of NH_4^+-N was significantly increased when nitrification was restricted. The concentrations of NH_4^+ at the SWI in the SP were 5.3 times than those in the NSP, which would mainly be governed by the thermal stratification, the seasonal changes of mineralization rate, and the water density in the deep reservoir [5,14], resulting in a stagnation of NH_4^+ in pore water and a higher nitrogen gradient during the SP. The diffusion fluxes of NO_3^- were presented as influxes from water towards the sediment, i.e., the NO_3^- was accumulated in the sediment in both the SP and NSP, which was mainly due to sufficient denitrification under anaerobic conditions, which significantly reduced the concentration gradient of NO_3^-. Compared with other reservoirs/lakes worldwide, the diffusion fluxes of NH_4^+-N at the SWI of the WJD reservoir were higher than those in Marano and Grado Lagoon in Italy, Danjiangkou Reservoir, Suma Park Reservoir, and Guanting Reservoir, which were less polluted, while the diffusion fluxes were closer to those at in the Yuqiao Reservoir, Upper Klamath Lake, and Erhai Lake (Table 6). This demonstrated that deep canyon reservoirs in Southwest China have relatively high NH_4^+ effluxes from the SWI, even though this region has been generally considered as a less disturbed and contaminated area, which implies that further studies are needed to assess the potential nitrogen pollution that can attributed to endogenous nitrogen release.

Table 6. Comparison of inorganic nitrogen diffusion fluxes at the SWI of other reservoirs.

Study Site	Location	Diffusion Flux (mg·m^{-2}·d^{-1})		Reference
		NH_4^+-N	NO_3^--N	
WJD Reservoir	China	9.48~15.66	−1.49~−0.21	This study
Danjiangkou Reservoir	China	0.39~17.66	−16.97~−4.33	[42]
Yuqiao Reservoir	China	4.38~30.57	−31.96~−4.13	[43]
Guanting Reservoir	China	1.59~13.00	-	[6]
Erhai Lake	China	8.97~74.84	-	[32]
Suma Park Reservoir	Australia	1.70 ± 1.20	0.30 ± 0.20	[20]
Marano and Grado Lagoon	Italy	4.88 ± 0.76	−21.30 ± 3.45	[44]
Upper Klamath Lake	America	4~134	−20~−0.1	[45]

Note: "-" was no data.

Yu et al. [46] investigated the sediment characteristics (including the sediment thickness and the concentrations of TN) of different sampling sites in the Dongfeng reservoir, a canyon reservoir along the same river as the one in this study, and they reported that although the sediment thickness varied spatially, the concentrations of TN in the surface sediment throughout the reservoir varied within a small range. Based on the updated study at the WJD reservoir [47]—and provided that the mean slope of hills was 30° and the mean depth was 90 m—the area of the surface sediment was estimated to be 11.6 km^2, and the total content of NH_4^+ in the water of the WJD reservoir reached 244 t. Thus, the total annual release of NH_4^+ was calculated to be 53.4 t, accounting for 21.8% of the total NH_4^+ content in the water of the WJD reservoir, which highlights the contribution of diffused

NH_4^+ from sediment to overlying water and the need for more attention in a future study to the evaluation of the environmental effects of endogenous nitrogen release in a high nitrogen-loading river basin.

5. Conclusions

This work aimed to identify the nitrogen transformation processes affected by ambient conditions of seasonal thermal stratification, to explore the factors controlling endogenous nitrogen release, and to quantify the nitrogen diffusion fluxes at the SWI of the WJD reservoir, an artificial deep reservoir in the Southeast Asian Karst Region. The vertical distribution profiles of nitrogen species verified that the mineralization rate of the upper layer (0–10 cm) overpassed that of bottom in the SP. Additionally, the nitrification that occurred in NSP illustrated the existence of NO_3^- at 0–19 cm in pore water. Even though similar phenomena have been observed in other lakes worldwide, the findings were seldom explained and reported in deep canyon reservoirs in Southwest China. We inferred that if thermal stratification commonly affected the nitrogen processes in most of the similar reservoirs, then the total NO_3^- influxes in deep reservoirs could be re-evaluated for a complete understanding of the nitrogen budget in impounded rivers worldwide. Combined with the seasonal distribution of δ^{15}N-PON and the C/N ratio, we also observed that the large isotope fractionation and significant seasonal differences in nitrogen species in the upper layer during the SP were mainly controlled by mineralization. Moreover, the sediment served as the source of NH_4^+-N and the reduction in NO_3^--N during the sampling time and the diffusion fluxes of sediment nitrogen were relatively close to those of seriously polluted lakes in China, which suggests that the deep canyon reservoirs may have an endogenous nitrogen transportation that is similar to that in polluted lakes, and that it may be enlarged by the water storage regulation mode that is used for better power generation. This study emphasized the influence of ambient conditions on endogenous nitrogen transformation and diffusion in a deep reservoir and provides an efficiently theoretical basis for managing watershed-scale nitrogen contamination in aquatic systems.

Author Contributions: Y.H.: methodology, writing—original draft preparation, and software; X.L.: conceptualization, investigation, data analysis, and manuscript editing; S.C.: sample collection, methodology, and investigation; J.R.: experimental investigation and sample collection; L.B.: data analysis and manuscript editing; J.L.: methodology and investigation; Y.G. and L.W.: sample collection and experimental investigation. All authors have read and agreed to the published version of the manuscript.

Funding: This research was funded by the National Natural Science Foundation of China (Grant Nos. 41661144029, 41672351) and the Natural Science Foundation of Tianjin City (18JCYBJC91000).

Institutional Review Board Statement: Not applicable.

Informed Consent Statement: Not applicable.

Data Availability Statement: The datasets used or analyzed during the current study are available from the corresponding author on reasonable request.

Conflicts of Interest: The authors declare no conflict of interest.

References

1. Green, P.A.; Vorosmarty, C.J.; Meybeck, M.; Galloway, J.N.; Peterson, B.J.; Boyer, E.W. Pre-industrial and contemporary fluxes of nitrogen through rivers: A global assessment based on typology. *Biogeochemistry* **2004**, *68*, 71–105. [CrossRef]
2. Valiela, I.; Geist, M.; McClelland, J.; Tomasky, G. Nitrogen loading from watersheds to estuaries: Verification of the waquoit bay nitrogen loading model. *Biogeochemistry* **2000**, *49*, 277–293. [CrossRef]
3. Zhang, Y.Q.; Wu, Z.H.; Xu, M.Y.; Pei, Z.L.; Lu, X.; Zhang, D.C.; Wu, T.; Li, B.; Xu, S.J. Nutrient deposition over the past 60 years in a reservoir within a medium-sized agricultural catchment. *Sci. Total Environ.* **2021**, *764*, 142896. [CrossRef]
4. Zeng, J.; Han, G.L. Tracing zinc sources with zn isotope of fluvial suspended particulate matter in Zhujiang river, Southwest China. *Ecol. Indic.* **2020**, *118*, 106723. [CrossRef]
5. Liu, X.L.; Liu, C.Q.; Li, S.L.; Wang, F.S.; Wang, B.L.; Wang, Z.L. Spatiotemporal variations of nitrous oxide (N_2O) emissions from two reservoirs in SW China. *Atmos. Environ.* **2011**, *45*, 5458–5468. [CrossRef]

6. Lei, P.; Zhu, J.J.; Zhong, H.; Pan, K.; Zhang, L.; Zhang, H. Distribution of nitrogen and phosphorus in pore water profiles and estimation of their diffusive fluxes and annual loads in guanting reservoir (Gtr), Northern China. *Bull. Environ. Contam. Toxicol.* **2021**, *106*, 10–17. [CrossRef] [PubMed]

7. Zhu, Y.Y.; Tang, W.Z.; Jin, X.; Shan, B.Q. Using biochar capping to reduce nitrogen release from sediments in eutrophic lakes. *Sci. Total Environ.* **2019**, *646*, 93–104. [CrossRef]

8. Aalto, S.L.; Saarenheimo, J.; Ropponen, J.; Juntunen, J.; Rissanen, A.J.; Tiirola, M. Sediment diffusion method improves wastewater nitrogen removal in the receiving lake sediments. *Water Res.* **2018**, *138*, 312–322. [CrossRef] [PubMed]

9. Hardison, A.K.; Algar, C.K.; Giblin, A.E.; Rich, J.J. Influence of organic carbon and nitrate loading on partitioning between dissimilatory nitrate reduction to ammonium (Dnra) and N_2 production. *Geochim. Et Cosmochim. Acta* **2015**, *164*, 146–160. [CrossRef]

10. Zarfl, C.; Lumsdon, A.E.; Berlekamp, J.; Tydecks, L.; Tockner, K. A global boom in hydropower dam construction. *Aquat. Sci.* **2015**, *77*, 161–170. [CrossRef]

11. Zeng, J.; Han, G.L.; Yang, K.H. Assessment and sources of heavy metals in suspended particulate matter in a tropical catchment, Northeast Thailand. *J. Clean. Prod.* **2020**, *265*, 10. [CrossRef]

12. Mulligan, M.V.; Soesbergen, A.; Saenz, L. GOODD, a global dataset of more than 38,000 georeferenced dams. *Sci. Data* **2020**, *7*, 1–8. [CrossRef]

13. Winton, R.S.; Calamita, E.; Wehrli, B. Reviews and syntheses: Dams, water quality and tropical reservoir stratification. *Biogeosciences* **2019**, *16*, 1657–1671. [CrossRef]

14. Dadi, T.; Rinke, K.; Friese, K. Trajectories of sediment-water interactions in reservoirs as a result of temperature and oxygen conditions. *Water* **2020**, *12*, 19. [CrossRef]

15. Muller, S.; Mitrovic, S.M.; Baldwin, D.S. Oxygen and dissolved organic carbon control release of n, p and fe from the sediments of a shallow, polymictic lake. *J. Soils Sediments* **2016**, *16*, 1109–1120. [CrossRef]

16. Liikanen, A.; Murtoniemi, T.; Tanskanen, H.; Vaisanen, T.; Martikainen, P.J. Effects of temperature and oxygen availability on greenhouse gas and nutrient dynamics in sediment of a eutrophic mid-boreal lake. *Biogeochemistry* **2002**, *59*, 269–286. [CrossRef]

17. Cai, Y.Y.; Cao, Y.J.; Tang, C.Y. Evidence for the primary role of phytoplankton on nitrogen cycle in a subtropical reservoir: Reflected by the stable isotope ratios of particulate nitrogen and total dissolved nitrogen. *Front. Microbiol.* **2019**, *10*, 2202. [CrossRef]

18. Ozkundakci, D.; Hamilton, D.P.; Gibbs, M.M. Hypolimnetic phosphorus and nitrogen dynamics in a small, eutrophic lake with a seasonally anoxic hypolimnion. *Hydrobiologia* **2011**, *661*, 5–20. [CrossRef]

19. Lavery, P.S.; Oldham, C.E.; Ghisalberti, M. The use of fick's first law for predicting porewater nutrient fluxes under diffusive conditions. *Hydrol. Process.* **2001**, *15*, 2435–2451. [CrossRef]

20. Chowdhury, M.; Al Bakri, D. Diffusive nutrient flux at the sediment-water interface in suma park reservoir, Australia. *Hydrol. Sci. J.* **2006**, *51*, 144–156. [CrossRef]

21. Liu, X.L.; Han, G.L.; Zeng, J.; Liu, M.; Li, X.Q.; Boeckx, P. Identifying the sources of nitrate contamination using a combined dual isotope, chemical and bayesian model approach in a tropical agricultural river: Case study in the mun river, Thailand. *Sci. Total Environ.* **2021**, *760*, 143938. [CrossRef]

22. Han, G.L.; Tang, Y.; Liu, M.; Van Zwieten, L.; Yang, X.M.; Yu, C.X.; Wang, H.L.; Song, Z.L. Carbon-nitrogen isotope coupling of soil organic matter in a karst region under land use change, Southwest China. *Agric. Ecosyst. Environ.* **2020**, *301*, 11. [CrossRef]

23. Zeng, J.; Yue, F.J.; Li, S.L.; Wang, Z.J.; Qin, C.Q.; Wu, Q.X.; Xu, S. Agriculture driven nitrogen wet deposition in a karst catchment in Southwest China. *Agric. Ecosyst. Environ.* **2020**, *294*, 10. [CrossRef]

24. Han, G.L.; Liu, C.Q. Water geochemistry controlled by carbonate dissolution: A study of the river waters draining karst-dominated terrain, Guizhou province, China. *Chem. Geol.* **2004**, *204*, 1–21. [CrossRef]

25. Zeng, J.; Han, G.L. Preliminary copper isotope study on particulate matter in Zhujiang river, Southwest China: Application for source identification. *Ecotoxicol. Environ. Saf.* **2020**, *198*, 8. [CrossRef] [PubMed]

26. Yin, R.; Wang, F.S.; Mei, Y.H.; Yao, C.Q.; Go, M.Y. Distribution of phosphorus forms in the sediments of cascade reservoir with different trophic states in Wujiang catchment. *Chin. J. Ecol.* **2010**, *29*, 91–97, (In Chinese with English Abstract).

27. Liu, X.L. Effects of Cascade Reservoirs Development on the Nitrogen Biogeochemical Processes in the Watershed—A Case Study of the Main Stream and Tributary in the Middle and Upper Reaches of the Wujiang River, Maotiao River. Ph.D. Thesis, Institute of Geochemistry, Chinese Academy of Science, Guiyang, China, 2010.

28. Hogarh, J.N.; Adu-Gyamfi, E.; Nukpezah, D.; Akoto, O.; Adu-Kumi, S. Contamination from mercury and other heavy metals in a mining district in Ghana: Discerning recent trends from sediment core analysis. *Environ. Syst. Res.* **2016**, *5*, 1–9. [CrossRef]

29. Lockhart, W.L.; Macdonald, R.W.; Outridge, P.M.; Wilkinson, P.; DeLaronde, J.B.; Rudd, J.W.M. Tests of the fidelity of lake sediment core records of mercury deposition to known histories of mercury contamination. *Sci. Total Environ.* **2000**, *260*, 171–180. [CrossRef]

30. Barra, R.; Cisternas, M.; Urrutia, R.; Pozo, K.; Pacheco, P.; Parra, O.; Focardi, S. First report on chlorinated pesticide deposition in a sediment core from a small lake in central chile. *Chemosphere* **2001**, *45*, 749–757. [CrossRef]

31. Copetti, D.; Tartari, G.; Valsecchi, L.; Salerno, F.; Viviano, G.; Mastroianni, D.; Yin, H.B.; Vigano, L. Phosphorus content in a deep river sediment core as a tracer of long-term (1962–2011) anthropogenic impacts: A lesson from the Milan metropolitan area. *Sci. Total Environ.* **2019**, *646*, 37–48. [CrossRef]

32. Zhao, H.C.; Zhang, L.; Wang, S.R.; Jiao, L.X. Features and influencing factors of nitrogen and phosphorus diffusive fluxes at the sediment-water interface of Erhai lake. *Environ. Sci. Pollut. Res.* **2018**, *25*, 1933–1942. [CrossRef]

33. Krom, M.D.; Berner, R.A. The diffusion coefficients of sulphate, ammonium and phosphate in anoxic marine sediments. *Limnol. Oceanogr* **1980**, *25*, 327–337. [CrossRef]

34. Ni, Z.X.; Zhang, L.; Yu, S.; Jiang, Z.J.; Zhang, J.P.; Wu, Y.C.; Zhao, C.Y.; Liu, S.L.; Zhou, C.H.; Huang, X.P. The porewater nutrient and heavy metal characteristics in sediment cores and their benthic fluxes in daya bay, South China. *Mar. Pollut. Bull.* **2017**, *124*, 547–554. [CrossRef]

35. Toussaint, E.; De Borger, E.; Braeckman, U.; De Backer, A.; Soetaert, K.; Vanaverbeke, J. Faunal and environmental drivers of carbon and nitrogen cycling along a permeability gradient in shallow North sea sediments. *Sci. Total Environ.* **2021**, *767*, 144994. [CrossRef] [PubMed]

36. Kraft, B.; Tegetmeyer, H.E.; Sharma, R.; Klotz, M.G.; Ferdelman, T.G.; Hettich, R.L.; Geelhoed, J.S.; Strous, M. The environmental controls that govern the end product of bacterial nitrate respiration. *Science* **2014**, *345*, 676–679. [CrossRef]

37. Liu, X.L.; Li, S.L.; Wang, Z.L.; Wang, B.L.; Han, G.L.; Wang, F.S.; Bai, L.; Xiao, M.; Yue, F.J.; Liu, C.Q. Sources and key processes controlling particulate organic nitrogen in impounded river-reservoir system on the Maotiao river, Southwest China. *Inland Waters* **2018**, *8*, 167–175. [CrossRef]

38. Liu, J.K.; Han, G.L. Tracing riverine particulate black carbon sources in Xijiang River Basin: Insight from stable isotopic composition and bayesian mixing model. *Water Res.* **2021**, *194*, 8. [CrossRef]

39. Yang, Y.X.; Xiang, P.; Lu, W.Q.; Wang, S.L. The sediment rate and burial fluxes of carbon and nitrogen in Wujiangdu reservoir, Guizhou, China. *Earth Environ.* **2017**, *45*, 66–73. (In Chinese with English Abstract)

40. Arndt, S.; Jorgensen, B.B.; LaRowe, D.E.; Middelburg, J.J.; Pancost, R.D.; Regnier, P. Quantifying the degradation of organic matter in marine sediments: A review and synthesis. *Earth-Sci. Rev.* **2013**, *123*, 53–86. [CrossRef]

41. Xiao, H.Y. Nitrogen Biogeochemical Cycles in a Seasonally Anoxic Lake. Ph.D. Thesis, Institute of Geochemistry, Chinese Academy of Science, Guiyang, China, 2002.

42. Wang, H.; Han, Y.P.; Pan, L.D. Spatial-temporal variation of nitrogen and diffusion flux across the water-sediment interface at the hydro-fluctuation belt of Danjiangkou reservoir in China. *Water Supply* **2020**, *20*, 1241–1252. [CrossRef]

43. Wen, S.L.; Wu, T.; Yang, J.; Jiang, X.; Zhong, J.C. Spatio-temporal variation in nutrient profiles and exchange fluxes at the sediment-water interface in Yuqiao reservoir, China. *Int. J. Environ. Res. Public Health* **2019**, *16*, 16. [CrossRef]

44. De Vittor, C.; Faganeli, J.; Emili, A.; Covelli, S.; Predonzani, S.; Acquavita, A. Benthic fluxes of oxygen, carbon and nutrients in the Marano and Grado lagoon (Northern Adriatic Sea, Italy). *Estuar. Coast. Shelf Sci.* **2012**, *113*, 57–70. [CrossRef]

45. Kuwabara, J.S.; Topping, B.R.; Lynch, D.D.; Carter, J.L.; Essaid, H.I. Benthic nutrient sources to hypereutrophic upper Klamath Lake, Oregon, USA. *Environ. Toxicol. Chem.* **2009**, *28*, 516–524. [CrossRef] [PubMed]

46. Yu, N.; Qin, Y.; Hao, F.; Lang, Y.; Wang, F. Using seismic surveys to investigate sediment distribution and to estimate burial fluxes of OC, N, and P in a Canyon reservoir. *Acta Geochim.* **2019**, *38*, 785–795. [CrossRef]

47. Chen, S.N.; Yue, F.J.; Liu, X.L.; Zhong, J.; Yi, Y.B.; Wang, W.F.; Qi, Y.L.; Xiao, H.Y.; Li, S.L. Seasonal variation of nitrogen biogeochemical processes constrained by nitrate dual isotopes in cascade reservoirs, Southwestern China. *Environ. Sci. Pollut. Res.* **2021**, *28*, 26617–26627. [CrossRef]

Article

Phosphorus Release from Sediments in a Raw Water Reservoir with Reduced Allochthonous Input

Bin Zhou [1,2,*], Xujin Fu [3], Ben Wu [2], Jia He [4], Rolf D. Vogt [5], Dan Yu [3], Fujun Yue [6] and Man Chai [3]

[1] State Key Laboratory of Environmental Criteria and Risk Assessment, Chinese Research Academy of Environmental Sciences, Beijing 100012, China

[2] Tianjin Academy of Eco-Environmental Sciences, Tianjin 300191, China; wuben@tj.gov.cn

[3] Tianjin Huanke Environmental Consulting Co., Ltd., Tianjin 300191, China; fuxujin2020@163.com (X.F.); yudan5303@163.com (D.Y.); achaim@126.com (M.C.)

[4] Beijing Key Laboratory of Urban Hydrological Cycle and Sponge City Technology, College of Water Sciences, Beijing Normal University, Beijing 100875, China; hejia@bnu.edu.cn

[5] Center for Biogeochemistry of the Anthropocene, Department of Chemistry, University of Oslo, N-0315 Oslo, Norway; r.d.vogt@kjemi.uio.no

[6] School of Earth System Science, Tianjin University, Tianjin 300072, China; fujun_yue@tju.edu.cn

* Correspondence: zhoubin19821214@163.com

Abstract: Following successful abatement of external nutrient sources, one must shift the focus to the role of phosphorus (P) release from sediment. This enables us to better assess the causes for sustained eutrophication in freshwater ecosystem and how to deal with this challenge. In this study, five sediment cores from the shallow YuQiao Reservoir in northern China were investigated. The reservoir serves as the main raw water source for tap water services of Tianjin megacity, with a population of 15.6 million. Sediment characteristics and P fractions were determined in order to assess the role of the sediments as the P source to the water body. The total P content (TP) in sediments was similar to what was found in catchment soils, although the P sorption capacity of sediments was 7–10 times greater than for the catchment soils. Isotherm adsorption experiments documented that when P concentration in overlying water drops below 0.032–0.070 mg L^{-1}, depending on the site, the sediment contributes with a positive flux of P to the overlying water. Adsorbed P at different depths in the sediments is found to be released with a similarly rapid release rate during the first 20 h, though chronic release was observed mainly from the top 30 cm of the sediment core. Dredging the top 30 cm layer of the sediments will decrease the level of soluble reactive phosphate in the water being sustained by the sediment flux of P.

Keywords: phosphorus release; internal P source; desorption P; abatement action; YuQiao Reservoir

Citation: Zhou, B.; Fu, X.; Wu, B.; He, J.; Vogt, R.D.; Yu, D.; Yue, F.; Chai, M. Phosphorus Release from Sediments in a Raw Water Reservoir with Reduced Allochthonous Input. *Water* **2021**, *13*, 1983. https://doi.org/10.3390/w13141983

Academic Editors: Guilin Han and Zhifang Xu

Received: 9 June 2021
Accepted: 16 July 2021
Published: 19 July 2021

Publisher's Note: MDPI stays neutral with regard to jurisdictional claims in published maps and institutional affiliations.

1. Introduction

1.1. Phosphorus Biogeochemistry and Indexes

Phosphorus (P) is an essential and usually limiting nutrient for primary production in aquatic ecosystems. Main anthropogenic P sources to surface water are generally distinguished between external sources, originating from diffuse (agricultural and aquaculture areas) or point sources (industrial and domestic sewage), and the internal source, which is P released from the sediments [1,2]. An increased flux of P to aquatic system degrades the water quality through eutrophication. The aquatic ecosystem responds to the increased nutrient levels with algae blooms pumping the P down into the lake sediments [3,4]. In the sediments, the mineralization of this organic P builds up a pool of adsorbed labile P. Depending on environmental conditions, the sediment may thereby also act as a P source, releasing labile P to the overlying water. Lake sediment is thus often the key element regulating the trophic level of its aquatic ecosystem [5]. A large number of biochemical and abiotic processes play important roles at the sediment–water interface governing the flux of P to or from the lake water [6,7]. Mineralized P in the sediment is in equilibrium with the pore water, which exchanges with the overlying water through diffusion and convection [8]. However, factors governing the rate of mineralization of P and the release mechanisms between the sediments and water are not well known.

As external pollution sources are eventually being increasingly curbed through effective abatement actions, the internal flux of P from the sediments is becoming the major P source, sustaining a high trophic level in many water bodies. Knowledge on how to limit this internal P loading from the sediment has thus become the key to achieve a sustainable restoration of lake ecosystems [1,8,9]. More studies on this issue are thus now required in order to enhance our ability to select optimum final abatement actions, solving the eutrophication challenge.

Risk of P release from sediments is mainly dependent on the form of P in the sediment, the concentration of P in the overlying water, the sediment texture and its pool of organic matter, as well as several physicochemical factors (i.e., temperature, pH, redox potential) [10–12]. The particle size distribution or texture of the sediment govern both the sorption capacity and the porosity, and thus the flow of pore-water. For example, high content of clay and organic matter in the sediments represented large surface area and number of sorption sites, facilitating a great P adsorption capacity [13].

Main forms of P in the sediments are calcium phosphates minerals P (Ca–P), non–mineral P (Fe/Al–P), exchangeable labile P (ex–P), and organic P (OP) [14]. Phosphorus sorption index (PSI) and degree of P saturation (DPS) are proxies that are commonly used to assess the risk of P release from sediments [15]. The threshold concentration of aqueous soluble reactive phosphorus (SRP) for P release is defined as the zero equilibrium P concentration (EPC$_0$) [14,16]. The difference between EPC$_0$ and SRP in overlying water is thus used as a criterion to determine whether the sediment acts as a source or sink of P., i.e., the sediment is a P source if the EPC$_0$ value is higher than SRP in the overlying water body. When deciding the depth of dredging, for removal of contaminated sediments, it is therefore necessary to relate the level of EPC$_0$ at the different depths in the sediments to the required level of SRP in the overlying water.

1.2. Shifting from External to Internal P Sources

YuQiao Reservoir (YQR) is the final water reservoir in the Luanhe–Tianjin water diversion development that supplies raw water to Tianjin for domestic supply, agricultural irrigation, and industrial use [17]. Thus, the water quality of the reservoir directly determines the safety of the city's water supply system. The rapid economic development within the Luanhe River and the reservoir's upper catchment basins has intensified the eutrophication in the reservoir since the 1990s [18,19]. Long–term monitoring data show that monthly TP concentrations and the annual tropic indexes have increased rapidly from 0.03 mg L^{-1} and 15.5 in 1990, to 0.04 mg L^{-1} and 35.5, respectively, in 2004 [20]. This posed a threat to drinking water quality and thus represented a potential risk for the water supply system of the Tianjin megacity. As the reservoir is the major drinking water source for Tianjin, its eutrophication has required increased investment for water treatment [20].

In the past, the omnipresence of agriculture in the vicinity around the reservoir and the presence of aquaculture in the reservoirs, as well as the disposal of sewage from the local population into channels, has contributed to a massive flux of nutrients and organic matter to the reservoir [21]. In the up–stream Daheiting Reservoir, the total reactive nitrogen (TN) concentration reached its maximum (5.3 mg L^{-1}) in 2013 and the TP concentration peaked at 0.34 mg L^{-1} in 2016 [22]. This eutrophication in the upstream reservoir led to a high frequency of algae blooms with anoxic conditions in the bottom water. This in turn posed a serious threat to the water quality of the downstream YQR. Owing to the rapid deterioration of upstream water and the release of endogenous phosphorous from the sediments of the reservoir, the water quality of YQR gradually deteriorated. The situation became acute in 2016 when a large–scale cyanobacteria bloom occurred in YQR with an algal density of more than 240 million cell L^{-1} [18]. This algae bloom resulted in a rapid decline in dissolved oxygen (around 0.7 mg L^{-1}), which seriously threatened the drinking water safety for the millions of citizens living in the Tianjin metropole, forcing the water supply from the YQR to be interrupted [22,23].

Summing up, external input was the main source for the high nutrient levels in the YQR water body. This external input comprised a number of sources, such as the water flux in the Luanhe River and the inflow from the Daheiting reservoir, as well as local seepage from numerous solid waste dumpsites, agricultural non–point sources, fishponds, and village sewage around the reservoir [20,24]. Since the massive alga bloom incident in 2016, a number of comprehensive abatement actions has successfully reduced the external flux of P to the reservoir. As a result, the external sources of nutrients to the YQR are now becoming gradually controlled [20]. The P flux from the sediments now constitute instead the major driving factor for the present sustained eutrophication of the water body [8]. Thus, investigation of its internal P sources with the aim of identifying targeted and optimal treatment measures, ensuring the abatement of algae blooms in the YQR, have become top priority for a sustainable management of the raw water source. In the present study, the contents of different P pools in sediment were determined, and laboratory simulation experiments of P release were conducted. The aim of this study is to: (1) investigate the spatial distribution of P forms in sediment cores; (2) assess factors governing release of P from the sediments; (3) quantify the rate of P release from the sediments; and (4) suggest suitable measure for abating the internal flux of P to the water body.

2. Materials and Methods

2.1. Study Area

YuQiao Reservoir (39°23′–40°23′ N; 117°26′–118°12′ E) is located in the northern part of Tianjin municipality. This reservoir was formed by a dam, impounded in the Zhou River in 1960, and has formed a shallow lake (0–12.7 m) with an average depth of 4.6 m (Figure 1). Through the Luanhe–Tianjin water diversion, the main water sources for YQR are the Panjiakou and Daheiting reservoirs. In addition, the Lin, Sha, and Li rivers (named Guo River after confluence, Figure 1) are three major tributaries that feed directly into the YQR. The entire catchment area of YQR is about 2060 km^2, with an average water storage capacity of 1.6×10^{10} m^3. The catchment is dominated by flat plains, with fluvo–aquic soils, comprising 34.6% of the catchment area. The hill (<500 m a.s.l.) and mountain (>500 m a.s.l.) regions, with mainly brown– and cinnamon soils, comprise 32% and 24.3%, respectively, of the catchment. Urban area accounts for the final 9.1% of the YQR basin area [21]. The reservoir water area in the plains is 74 km^2, accounting for 3.6% of the basin area [12,21]. Farmland on the plains and forest in the hills are the two main land uses, covering 33.0% and 40.9% of the entire catchment, respectively. Paddy fields account for 12.6% of the farmland area. Annual average temperature and rainfall is between 10.4–11.5 °C and 748.5 mm, respectively. The rainfall shows seasonal monsoonal variation, with most precipitation during the summer months July and August.

2.2. Sampling and Analysis

Five sediment cores were collected from the YQR, capturing the range of water depth (Figure 1) in April 2017. To decrease the direct impact from tributaries, the distance between the sediment core sampling sites and the fluctuation zone is more than 1.5 km, e.g., YQ 2. Water depth increases in the direction of main flow from east (YQ 1) to west (YQ 4), with the deepest water level close to the dam (YQ 5). Sediment cores were sectioned in different depths. For example, each sediment core was sectioned into 10 cm layers along the top 0–50 cm, and at 15 cm intervals below 50 cm for soil basic parameters analysis, while the sectioning into 30 cm layers was used for simulation experiments. All sediment samples were sealed in labeled centrifuge tubes and transported to the laboratory on ice. Sediment pH, particle size distribution (PSD), organic matter content, and effective cation exchange capacity (CECe) were determined using methods as described in a preceding study [25]. Briefly, PSD between silt (0.002–0.05 mm), very fine sand (0.05–0.1 mm), fine sand (0.1–0.25 mm), and clay (<0.002) were determined by a laser particle size analyzer. Organic matter content was determined gravimetrically as the loss on ignition (LOI) at 550 °C, and CECe was measured by extraction using $BaCl_2$. Amorphous Fe and Al–oxyhydroxide were extracted separately with ammonium oxalate (Fe_{ox} and Al_{ox}) and citrate–bicarbonate–dithionite (Fe_{cbd}, Al_{cbd}), and determined in the extracts using ICP–OES. All of the sediment characteristics were determined for all core sections, except for the YQ 1 sediment core, owing to limited samples.

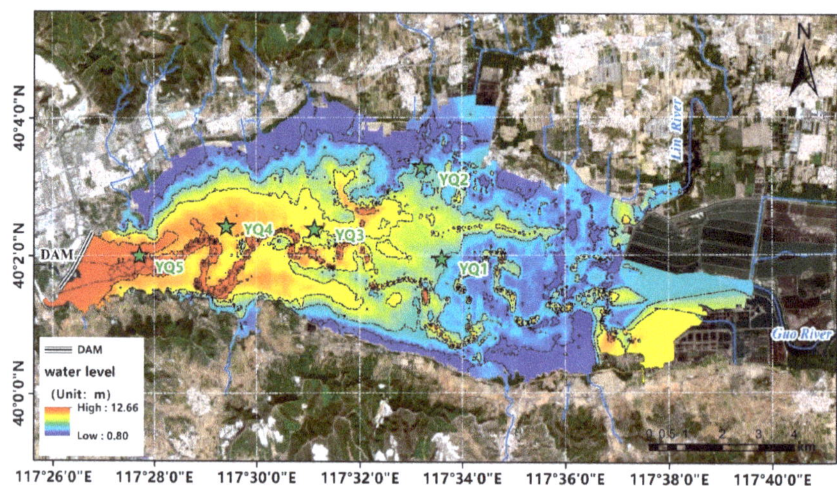

Figure 1. YuQiao Reservoir marked with color–coded water depth and location of sampling sites (YQ 1–5). The light brown and green colors in the surroundings area are mainly agricultural area and mountains, respectively.

Total inorganic P (IP) in the sediments was extracted using hot (70 °C) 6 mol L^{-1} H_2SO_4. The total P (TP) content was determined in a similar manner after combustion of the sediment sample at 550 °C, while organically bound P (OP) was calculated as the difference between TP and IP. The P in extracts of TP and IP were analyzed using Inductively Coupled Plasma Optic Emission Spectroscopy (ICP–OES). Soil test phosphorus concentration (STP), commonly used for agricultural nutrient management, was conducted using the Mehlich–3 extractant [26]. STP in the extract was determined by the molybdenum blue method [27]. P pools in sediment core segments were sequentially extracted into exchangeable P (EX–P), P adsorbed to amorphous iron (Fe–P), and calcium phosphate minerals P (Ca–P), using $MgCl_2$, $NaOH + NaCO_3$, and $HAc + NaAc$ as extractants, respectively [28].

2.3. Sorption Index and P Saturation

P sorption index (PSI) [15] is determined as a single–point isotherm by mixing the sediment with an excess of orthophosphate–P (1.5 mg g^{-1} sediment) and measuring the concentration of bioavailable P (i.e., STP) remaining in solution. The index is given as the ratio of sorbed (X in mg P kg^{-1}) over the log concentration of STP (log C in mg P L^{-1}) according to Equation (1):

$$PSI\ (L/kg) = \frac{X}{\log C} \tag{1}$$

Bache and Williams (1971) found that both the arbitrary value of PSI and STP provided suitable relative proxies for the P sorption capacity (PSC) of the soil. The sum of PSI and STP is therefore applied as a measure for soil PSC in this study. Degree of P Saturation (DPS) in the sediments was accordingly determined using Equation (2).

$$\text{DPS (\%)} = \frac{\text{STP}}{\text{PSC}} \times 100\% = \frac{\text{STP}}{(\text{STP} + \text{PSI})} \times 100\% \tag{2}$$

2.4. Sorption Experiments Fordetermination of EPC_0

Aliquots of 0.5 g integrated sediment samples from 0–30 cm and 30–60 cm of the cores were mixed with 50 mL solution of different KH_2PO_4 concentrations (0.00, 0.025, 0.05, 0.1, 0.2, 0.5, 1, 5, 15, 50, 75 mg L^{-1}) in centrifuge tubes at room temperature. The samples were shaken for 24 h on a shaker set to rotate at 260 rpm. The supernatant was separated by centrifugation immediately after sampling and filtered through a 0.45 μm membrane filter. Phosphate in the filtrate was analyzed using the molybdenum blue method.

2.5. Phosphorus Release Experiment

Each sediment core was sectioned at 30 cm intervals (0–30, 30–60, and 60–90 cm) into three sediment columns. The level of filtered ambient lake water, which was collected from one site, above the open sediment column was kept between 60–80 cm. After 1, 2, 8, 15, 30, 60, 120, 180, 240 h, 0.1 L water was gently collected at 10 cm above sediment surface and replaced by filtered lake sample water, avoiding any disturbance of the sediment.

The release rate of P (R) is calculated using Equation (3) [22]:

$$R = [V \times (Cn - C0) + \sum_{j=1}^{n} V_{j-1} \times (C_{j-1} - C_a)]/S \times t \tag{3}$$

where R is release rate (mmol m^{-2} d^{-1}); V is the water volume in the sedimentary column (L); C_0, C_n, C_{j-1} are SRP concentration (mmol P L^{-1}) in the samples collected at the initial, n, and $j-1$ times, respectively. C_a is the SRP concentration of added water sample (mmol L^{-1}); V_{j-1} is the volume of collected water sample at time $j-1$; S is the interface surface area of water and sediment (m^2); and t is the release time (d). SRP was determined by the molybdenum blue method.

2.6. Data Statistics

SPSS 22.0 and Excel 2019 were used for statistical data analysis. Origin2017C software (Originlab, Northampton, MA, USA) was used for figure drawing. Correlation between variables were analyzed by Pearson's correlation coefficients ($p < 0.05$).

3. Results and Discussion

3.1. Sediment Core Characteristics and Variablity

The sediment in YQR is silt loam mainly composed of silt with a minor fraction of clay (Figure 2). This is similar to the particle size distribution of soils in the flat plain region of the local catchment [29]. The clay content decreased further from the top to the bottom of sediment core (Figure 2).

The organic matter content in the sediments is also low, with LOI ranging only from 0.62 to 4.9%, with a mean value of only 1.9% (Table 1, Figure 1). Still, this is 60% higher than what was measured at the end of the 1980s (1.198%) [30]. The highest organic content is found in the top sediments (0–30 cm) (Figure 3), especially in the two middle sediment cores (YQ 3 and 4). The low average LOI are found at the shallowest (YQ1) and deepest water depth (YQ 5) of the reservoir (Table 1, Figure 1). The organic matter content decreases with sediment depth, reaching an organic content in the sediments deeper than 80 cm, which is close to the levels that were measured at the end at the 1980s [30]. Although this is likely partly due to the gradual mineralization of the organic matter in the sediments over time, it may also reflect a rapid enrichment of deposited organic matter over the past 30 years. Based on the comparably low clay and organic matter content, the sorption capacity at the sediments is expected to be low, and the risk of phosphate release is thus high.

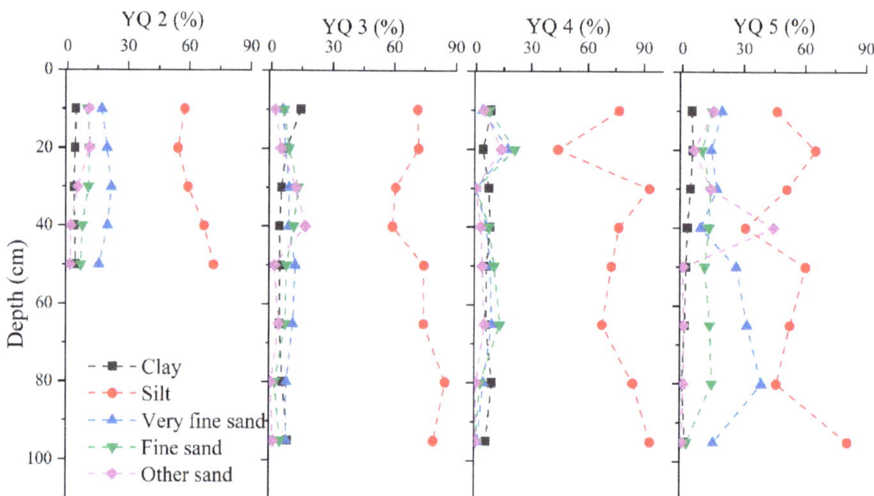

Figure 2. The vertical variation of particle size distribution at four sites.

CECe reflect the sediment's ability to hold cations, which influences other soil chemical properties, e.g., adsorption and desorption capacity, and pH [25,31]. The CECe of sediment cores ranged from 180 to 379 mmol/kg, with a mean value of 294 mmol/kg (Figure 3). The base saturation is high with calcium and magnesium ions contributing more than 85% of exchangeable cations, while the percentage of exchangeable iron and aluminum ions are relatively minor. The lowest average CECe are found at YQ 2 and 4, while the highest average CECe are observed at YQ 3 and 5 (Table 1). The spatial variation in CEC_e does thus not appear to be governed by the clay nor organic matter content.

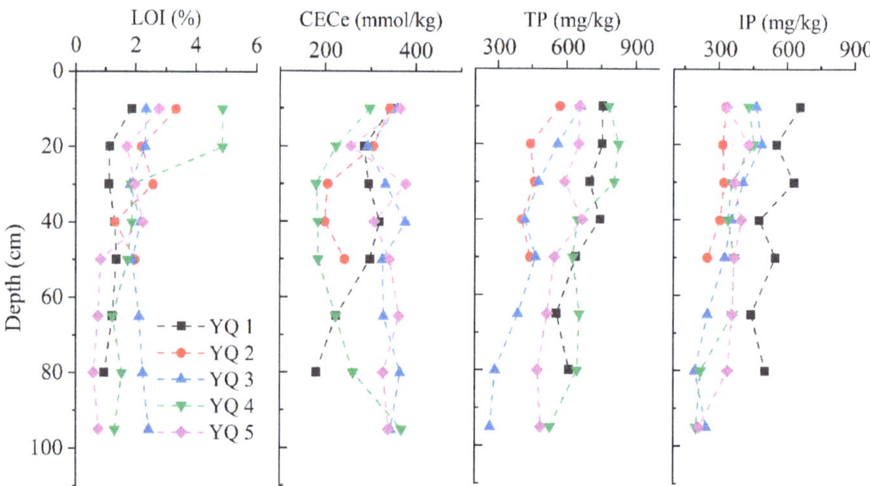

Figure 3. The vertical variation of LOI, CECe, TP, and IP at four sites.

Table 1. Characteristics of five sediment cores in YuQiao Reservoir.

Core Depth (cm) Subsections	YQ 1 80 7		YQ 2 50 5		YQ 3 95 8		YQ 4 95 8		YQ 5 95 8	
	Avg	CV [a]	Avg	CV [a]	Avg	CV [a]	Avg	CV [a]	Avg	CV [a]
pH	6.5	6	7.6	1	6.9	3	6.9	3	7.2	3
Clay [b]	–	–	3.85	13	6.58	53	6.5	25	2.58	69
LOI [b]	1.28	22	2.29	33	2.17	10	2.42	64	1.47	56
CECe [c]	278	20	259	24	341	8	240	28	335	12
TP [d]	681	12	465	14	442	30	691	15	575	14
IP [d]	545	14	306	11	342	31	341	26	352	18
OP [d]	136	52	159	36	99	56	352	19	223	28
STP [d]	–	–	16.1	73	21.5	72	14.0	71	20.2	58
Ex–P [d]	–	–	3.9	30	7.3	31	4.6	30	9.5	39
Fe–P [d]	–	–	7.1	40	50.4	58	106	53	82.3	16
Ca–P [d]	–	–	164	12	187	31	194	29	247	17
PSI [d]	–	–	1135	14	1106	17	1424	19	1338	7
PSC [d]	–	–	1151	14	1128	17	1438	19	1358	6

The units of a, b, c, and d are %, %, mmol/kg, and mg/kg, respectively. Avg and CV mean the average value and coefficient variation.

The concentration of TP in the sediment ranged from 267 to 827 mg/kg, with an average of 577 mg/kg (Figure 3). This is similar to what was found in the local catchment soil [25]. Compared to the levels measured in the 1980s (440 mg/kg) the TP has increased by nearly 30% [32]. It is worth noting that the TP in the middle of the reservoir (YQ 4 with the highest organic content in the surface layers) reached 827 mg/kg at 10–20 cm (Figure 3). Along with the clay and organic matter content, the TP also gradually decreased with depth, but the overall decrease was less than 20%. According to the sediment evaluation standard, formulated by the Ontario Ministry of Environment and Energy (1992), TP concentration in freshwater sediments exceeding 600 mg/kg will cause eutrophication. Although three of the sediment cores with average TP concentrations contain less than 600 mg kg^{-1} (Table 1), the top 10 cm of sediments in four sediment cores (YQ 1, 3–5) had a TP concentration that exceeded this safety threshold (Figure 3).

IP dominated significantly over OP as the main P pool (mean value 67%, Figure 3). The exception is for the sediment core with the highest organic matter content and TP (YQ 4), which has similar OP and IP contribution. Since IP is the main P fraction in the sediment cores, this study mainly focused on the pools of inorganic phosphorus compounds, including Ex–P, Fe–P, and Ca–P. The results show that Ca–P accounted for 20 to 75% of total phosphorus, while Fe–P accounted for 0.4 to 30% of total phosphorus (Table 1). Ca-P is mainly constituted by calcium phosphate minerals, reflecting a high concentration of calcium ions in solution from weathering of carbonates [33]. The release of phosphate from Fe-P is, on the other hand, governed by the redox potential. As the main active form of P in sediments, Fe–P transformation at the sediment–water interface mainly depends on the fluctuating redox potential to effectively control the entire process of sediment phosphorus adsorption and desorption [34,35]. Under oxygen–rich conditions, Fe^{3+} and P combine to form the rather insoluble ferric phosphate (Fe^{3+}–P), which is stored in the sediments. Under anoxic conditions, the Fe^{3+} may act as an electron acceptor in the heterotrophic oxidation of organic matter. This reduces it to Fe^{2+}, which forms the much more soluble ferrous phosphate, allowing the iron–bound phosphorus to be released to the pore water. The labile phosphorus in the sediments is therefore mainly contributed by the Fe-P pool. This is causing the variations in the release of P in sediments to be mainly governed by changes in the redox state and thereby the valence state of iron [36]. Since bioavailable phosphorus is readily assimilated by the biota, the dissolved labile and bioavailable phosphorus content and its percentage of TP are held low despite a large influx [5].

During the summer season in eutrophic waters, there is a high primary production in the surface water layer with photosynthetically active radiation (PAR). This leads to a rich flow of organic matter drizzling down to the sediment, where it becomes partly decomposed and mineralized through heterotrophic respiration. This consumes dissolved oxygen and causes hypoxia below the PAR zone [31,37]. In addition, higher temperatures during the summer increase the activity of microorganisms in the sediment and lower the overall capacity of the water to hold oxygen, causing a more rapid loss of oxygen in the water. During the summer season, due to the rapid accumulation primary productivity in surface water, the combing of low dissolved oxygen at the bottom of the water used by the degradation and high pH quickly reduced Fe^{3+} to Fe^{2+} [1,38]. The anoxic conditions then cause the Fe^{3+} to be reduced to Fe^{2+}, releasing P into solution. This also indirectly implies that by sustaining a high dissolved oxygen level in the water column, it is possible to inhibit the release of P from the sediments and thereby obtain and sustain low phosphorus levels. Under strong alkali conditions (pH > 9), a further increase in pH may cause increased phosphorus release from the sediment [11]. A preceding study of the YQR found that the pH of the water was higher than 9 (i.e., between 9.11 and 10.16) [8]. These high pH levels, which especially arise during alga blooms in the summer season, are likely to diffuse into the pore water of the sediment. The high concentrations of OH^- will replace the adsorbed phosphate, resulting in a release of phosphate to the pore water. The Fe–P content of the sediment is therefore an important indicator of the potential of phosphorus release from the sediment [36,39].

The STP concentration ranged from 3.5 to 49.5 mg/kg, with a mean value of 18.5 mg/kg. This was lower than what is found in the catchment soil [25]. However, the PSI values were 7–10 times higher in the sediment than in the catchment soil. This is indicating that the sediments have much higher P sorption capacities than the catchment soil. Compared to catchment soil, the reservoir sediments are thus more prone to adsorb P. On the other hand, as the sediments are typically in an anaerobic state and are exposed to alkaline pH, some of the phosphate bound to Fe^{3+}–P in the sediments would be released to the overlying water. DPS is an indicator to evaluate the degree of P saturation and is used to predict the potential risk for P release, i.e., low DPS value implies a high potential risk of P release. The DPS of sediment cores, ranging from 0.24 to 2.91%, are lower than for the catchment soil [25]. There is therefore a greater risk of P loss from the sediments than from the catchment soils.

3.2. Governing Factors for Adsorption and Desorption of P in the Sediments

To understand the effects of sediment texture and pools of P on P release, the correlations between PSI and various explanatory factors were assessed in the four fully analyzed sediment cores (YQ 2–5, Figure 4). STP is negatively correlated with organic matter content (LOI), Fe_{ox} and Al_{ox}, Fe_{cdb}, and Al_{cdb} concentrations. This reflects that the cocktail of chemicals in the Mehlich–3 extractant is not able to release the P that is bound to organic matter, as well as Fe and Al, despite that the fluoride in the Mehlich–3 is supposed to extract iron and aluminum phosphates. The higher content of organic matter, amorphous iron, and aluminum oxyhydroxides in the sediments, the less is extracted as STP. The Ca–P is positively correlated with fine sand and negatively correlated with silt. This is indicating that the levels of Ca–P in the sediments is mainly reflected by sediment texture, possibly because the calcium phosphate minerals are mainly present as fine–sand–sized particles. The TP concentration is positively correlated to the amount of Al oxyhydroxides and amorphous Fe. The risk of P release (PSI) is significantly ($p < 0.01$) correlated to the IP concentration, as well as the Al and Fe concentrations. Following the discussion above, it is worth noticing that pH is found to be significantly ($p < 0.01$) correlated to Fe–P concentration in the sediments (Figure 4). This may be due to the fact that sediments with high pH may have had increased desorption of phosphate from the Fe–P pool in the sediment to the overlying water.

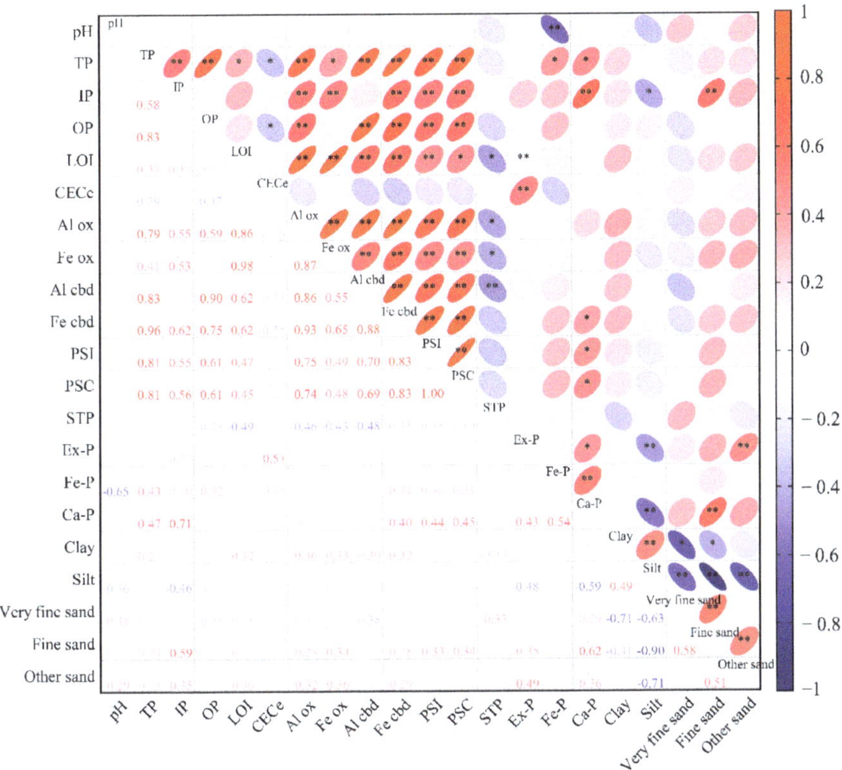

Figure 4. Correlation analysis among variables in the YQ 2–5 sediment cores. * Correlation is significant at $p < 0.05$; ** Correlation is significant at $p < 0.01$.

3.3. Phosphorus Adsorption Isotherms

Sediment cores from the north of the reservoir (YQ 2), middle of reservoir (YQ 3), and in front of the dam (YQ 5) were selected for the phosphorus isotherm adsorption analysis. With the exception of 2016 [22], the TP in the reservoir over the past 10 years has remained below 0.05 mg L^{-1} and the concentration of soluble orthophosphate has been less than 0.01 mg L^{-1}. Now that the external point and non–point sources have been successfully abated, the sediments are activated as an internal source of P, sustaining a high trophic status of the water. In 2016, the annual average TP concentration of the reservoir exceeded 0.32 mg L^{-1}, and the concentration of soluble orthophosphate exceeded 0.14 mg L^{-1} [22]. During this period, the sediments accumulated the detrital organic matter and adsorbed a great amount of P from the water.

The EPC$_0$ values decreased clearly with increasing sediment depth in YQ 3 and YQ 4 (Figure 5), indicating that the risk of P release is lower in the deeper sediment layers. The shift from adsorption to desorption (EPC$_0$) occurred at SRP levels between 0.032–0.070 mg L^{-1} (except for YQ 2, 30–60 cm, Figure 5), with an average concentration close to 0.05 mg L^{-1}. That means that the phosphorus in the sediment will release as an internal source when the SRP concentration in the water is lower than 0.05 mg L^{-1}. Considering the high EPC$_0$ levels in the top 30 cm of the sediment, the sediments of YQR have been releasing P to the water during this past decade. According to the results in YQ 3 and 5, the removal of the top 30 cm of the sediments will lower the STP concentration in the water to between 0.032 and 0.039 mg P L^{-1} (30–60 cm, Figure 5), at which the sediments start to release STP to the water.

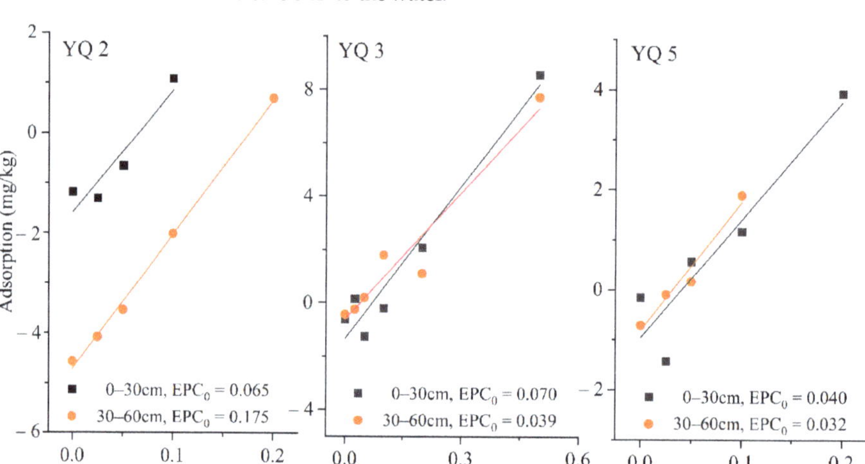

Figure 5. Phosphorus adsorption isotherms in sediment cores at two different sediment depths. Please note different scales on the Y–axis.

3.4. Phosphorus Release Rate from Sediment Cores

The release of P from the top (0–30 cm), middle (30–60 cm), and bottom (60–90 cm) of the sediment cores to the ambient lake water were measured by column experiments. Time series of the P levels in the water above the sediment cores (Figure 6) show that P is released at all three sediment depths. Rapid release was observed during the first 20 h. After 20 h, the top sediment sustained a higher release than that of the middle and bottom core sections at YQ 2, 3, and 5. This is likely due to the fact that the top sediment cores have a higher content of TP and IP (Figure 3). The exception is again YQ 4, which had the highest release at the middle depth (Figure 6). After 6 days, reflecting a long–term release, the sediments still sustained a significant impact on the phosphorus concentration in the overlying water. In the absence of any external source input, the 0–30 cm top sediments at the four sites were able to maintain an elevated SRP concentration level in the overlying water, e.g., higher than 0.05 mg L^{-1} at the YQ 3 (Figure 7). The middle and bottom sections of the sediment cores maintained a lower phosphorus concentration between 0.02–0.04 mg L^{-1}.

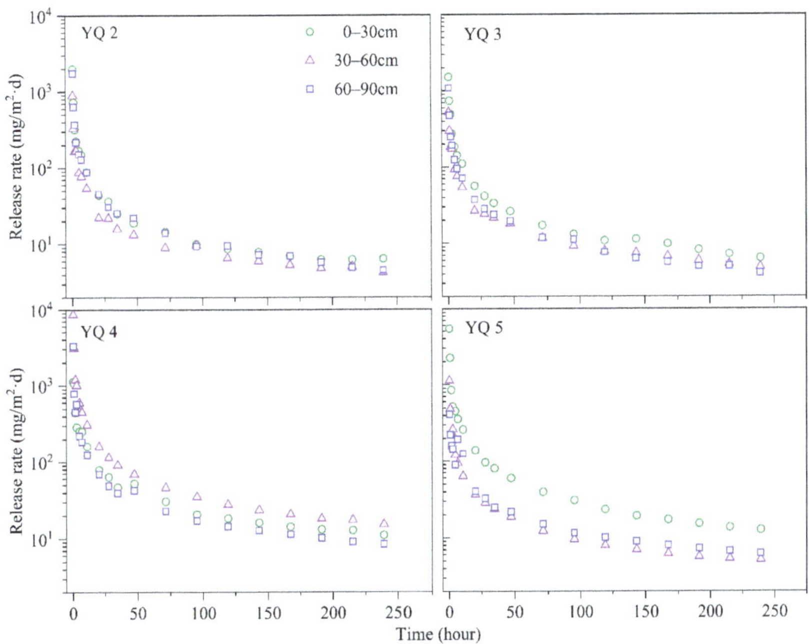

Figure 6. Time series of P release rate from four sediment cores and three different sediment depths.

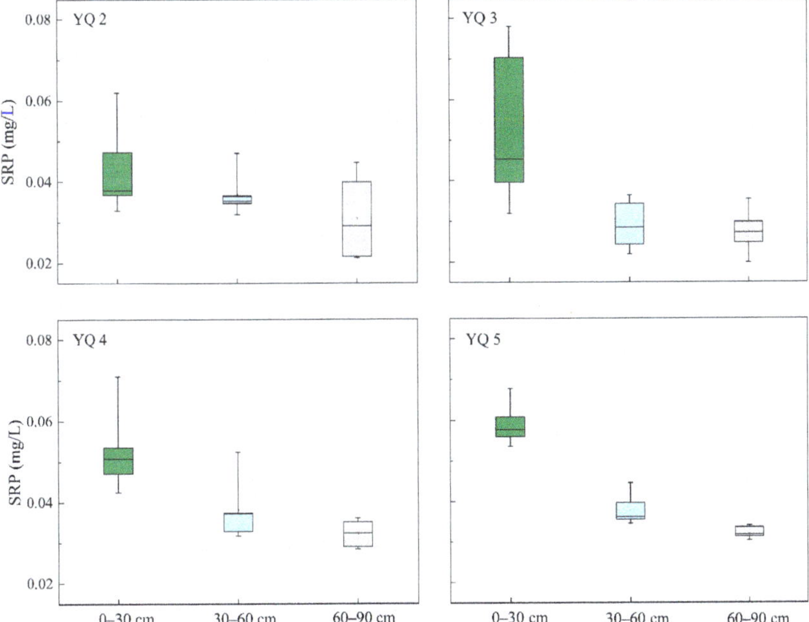

Figure 7. Boxplot for SRP concentration in ambient lake water above the sediments during the simulation of sediment release. The edges of the boxplot represent the 75th and 25th percentiles, respectively. The solid line in the box represented the median value. The branch gave the range of the data except for the outliers.

4. Conclusions

The sediments in the YuQiao drinking water reservoir in Tianjin, China, have accumulated high levels of phosphorus. Now that the external anthropogenic sources of phosphorous to the reservoir are reduced, the phosphorus release from the sediment to the water has increased. The large pool of phosphorus in the sediments will continue to produce a flux of P into the water body, sustaining elevated levels of SRP, causing annual algal blooms. Alga blooms during summer may result in a rapid increase in the pH value, enhancing the replacement effect of the large pool of adsorbed phosphate and causing further eutrophication. Moreover, increased summer temperatures strengthen the anoxic conditions, thereby accelerating the release of phosphorus bound to ferric iron from the sediment. Without abatement actions specifically targeted to reduce or completely remove this internal phosphorous source, it will not be possible to restore the water quality in the YuQiao Reservoir to good ecological conditions. The areas in front of the dam (YQ 5), as well as the middle of the reservoir sediment core (YQ 3), are high–risk areas of phosphorus release that need to be given priority attention in future governance. Our study shows that by removing the top 30 cm of the sediments, the levels of SRP in the water will be reduced to below 0.04 mg P L^{-1}.

Author Contributions: B.Z.: Investigation, data curation, formal analysis, writing original draft, project administration, and funding acquisition. X.F.: Formal analysis, methodology, investigation, review, and editing. B.W. and J.H.: Investigation, data curation, and review. D.Y. and M.C.: Investigation, data curation, and review. F.Y.: review and editing. R.D.V.: Conceptualization, review, and editing. All authors have read and agreed to the published version of the manuscript.

Funding: This research was supported by Tianjin Science and Technology Project (Grant No. 18ZXSZSF00130).

Institutional Review Board Statement: Not applicable.

Informed Consent Statement: Not applicable.

Data Availability Statement: The data presented in this study is available on request from the corresponding author.

Conflicts of Interest: The authors declare no conflict of interest.

References

1. Chen, Q.; Chen, J.G.; Wang, J.F.; Guo, J.Y.; Jin, Z.X.; Yu, P.P.; Ma, Z.Z. In situ, high–resolution evidence of phosphorus release from sediments controlled by the reductive dissolution of iron–bound phosphorus in a deep reservoir, southwestern China. *Sci. Total Environ.* **2019**, *666*, 39–45. [CrossRef] [PubMed]
2. Nowlin, W.H.; Evarts, J.L.; Vanni, M.J. Release rates and potential fates of nitrogen and phosphorus from sediments in a eutrophic reservoir. *Freshw. Biol.* **2005**, *50*, 301–322. [CrossRef]
3. Cerco, C.F.; Noel, M.R. Impact of Reservoir Sediment Scour on Water Quality in a Downstream Estuary. *J. Environ. Qual.* **2016**, *45*, 894–905. [CrossRef] [PubMed]
4. Maavara, T.; Parsons, C.T.; Ridenour, C.; Stojanovic, S.; Durr, H.H.; Powley, H.R.; Van Cappellen, P. Global phosphorus retention by river damming. *Proc. Natl. Acad. Sci. USA* **2015**, *112*, 15603–15608. [CrossRef] [PubMed]
5. Sondergaard, M.; Jensen, J.P.; Jeppesen, E. Role of sediment and internal loading of phosphorus in shallow lakes. *Hydrobiologia* **2003**, *506*, 135–145. [CrossRef]
6. Corzo, A.; Jiménez–Arias, J.L.; Torres, E.; García-Robledo, E.; Lara, M.; Papaspyrou, S. Biogeochemical changes at the sediment–water interface during redox transitions in an acidic reservoir: Exchange of protons, acidity and electron donors and acceptors. *Biogeochemistry* **2018**, *139*, 241–260. [CrossRef]
7. Santschi, P.; Höhener, P.; Benoit, G.; Buchholtz-ten Brink, M. Chemical processes at the sediment-water interface. *Mar. Chem.* **1990**, *30*, 269–315. [CrossRef]
8. Wen, S.L.; Wang, H.W.; Wu, T.; Yang, J.; Jiang, X.; Zhong, J.C. Vertical profiles of phosphorus fractions in the sediment in a chain of reservoirs in North China: Implications for pollution source, bioavailability, and eutrophication. *Sci. Total Environ.* **2020**, *704*, 135318. [CrossRef]
9. Zeng, J.; Han, G. Preliminary copper isotope study on particulate matter in Zhujiang River, southwest China: Application for source identification. *Ecotoxicol. Environ. Saf.* **2020**, *198*, 110663. [CrossRef] [PubMed]
10. Boström, B.; Andersen, J.M.; Fleischer, S.; Jansson, M. Exchange of phosphorus across the sediment-water interface. *Hydrobiologia* **1988**, *170*, 229–244. [CrossRef]

11. Wu, Y.H.; Wen, Y.J.; Zhou, J.X.; Wu, Y.Y. Phosphorus release from lake sediments: Effects of pH, temperature and dissolved oxygen. *KSCE J. Civ. Eng.* **2014**, *18*, 323–329. [CrossRef]

12. Zhang, C.; Zhang, W.N.; Huang, Y.X.; Gao, X.P. Analysing the correlations of long-term seasonal water quality parameters, suspended solids and total dissolved solids in a shallow reservoir with meteorological factors. *Environ. Sci. Pollut. Res.* **2017**, *24*, 6746–6756. [CrossRef] [PubMed]

13. Gérard, F. Clay minerals, iron/aluminum oxides, and their contribution to phosphate sorption in soils—A myth revisited. *Geoderma* **2016**, *262*, 213–226. [CrossRef]

14. Bao, L.L.; Li, X.Y.; Su, J.J. Alteration in the potential of sediment phosphorus release along series of rubber dams in a typical urban landscape river. *Sci. Rep.* **2020**, *10*, 2714. [CrossRef] [PubMed]

15. Bache, B.W.; Williams, E.G. A PHOSPHATE SORPTION INDEX FOR SOILS. *J. Soil Sci.* **1971**, *22*, 289–301. [CrossRef]

16. Cao, X.Y.; Chen, X.Y.; Song, C.L.; Zhou, Y.Y. Comparison of phosphorus sorption characteristics in the soils of riparian buffer strips with different land use patterns and distances from the shoreline around Lake Chaohu. *J. Soils Sediments* **2019**, *19*, 2322–2329. [CrossRef]

17. Zhai, H.Y.; He, X.Z.; Zhang, Y.; Du, T.T.; Adeleye, A.S.; Li, Y. Disinfection byproduct formation in drinking water sources: A case study of Yuqiao reservoir. *Chemosphere* **2017**, *181*, 224–231. [CrossRef] [PubMed]

18. Huo, D.; Chen, Y.X.; Zheng, T.; Liu, X.; Zhang, X.Y.; Yu, G.L.; Qiao, Z.Y.; Li, R.H. Characterization of Microcystis (Cyanobacteria) Genotypes Based on the Internal Transcribed Spacer Region of rRNA by Next-Generation Sequencing. *Front. Microbiol.* **2018**, *9*, 971. [CrossRef] [PubMed]

19. Zhou, B.; Vogt, R.D.; Xu, C.Y.; Lu, X.Q.; Xu, H.L.; Bishnu, J.P.; Zhu, L. Establishment and Validation of an Amended Phosphorus Index: Refined Phosphorus Loss Assessment of an Agriculture Watershed in Northern China. *Water Air Soil Pollut.* **2014**, *225*, 2103. [CrossRef]

20. Li, X.; Xu, Y.; Zhao, G.; Shi, C.L.; Wang, Z.L.; Wang, Y.Q. Assessing threshold values for eutrophication management using Bayesian method in Yucliao Reservoir, North China. *Environ. Monit. Assess.* **2015**, *187*, 195. [CrossRef] [PubMed]

21. Zhou, B.; Xu, Y.P.; Vogt, R.D.; Lu, X.Q.; Li, X.M.; Deng, X.W.; Yue, A.; Zhu, L. Effects of Land Use Change on Phosphorus Levels in Surface Waters-a Case Study of a Watershed Strongly Influenced by Agriculture. *Water Air Soil Pollut.* **2016**, *227*, 160. [CrossRef]

22. Wen, S.L.; Wu, T.; Yang, J.; Jiang, X.; Zhong, J.C. Spatio-Temporal Variation in Nutrient Profiles and Exchange Fluxes at the Sediment-Water Interface in Yuqiao Reservoir, China. *Int. J. Environ. Res. Public Health* **2019**, *16*, 3071. [CrossRef]

23. Chen, Y.Y.; Zhang, C.; Gao, X.P.; Wang, L.Y. Long-term variations of water quality in a reservoir in China. *Water Sci. Techol.* **2012**, *65*, 1454–1460. [CrossRef]

24. Chang, C.; Sun, D.M.; Feng, P.; Zhang, M.; Ge, N. Impacts of Nonpoint Source Pollution on Water Quality in the Yuqiao Reservoir. *Environ. Eng. Sci.* **2017**, *34*, 418–432. [CrossRef]

25. Zhou, B.; Vogt, R.D.; Lu, X.Q.; Yang, X.G.; Lu, C.W.; Mohr, C.W.; Zhu, L. Land use as an explanatory factor for potential phosphorus loss risk, assessed by P indices and their governing parameters. *Environ. Sci. Process. Impacts* **2015**, *17*, 1443–1454. [CrossRef] [PubMed]

26. Sharpley, A.N. Dependence of Runoff Phosphorus on Extractable Soil Phosphorus. *J. Environ. Qual.* **1995**, *24*, 920–926. [CrossRef]

27. Asher, L.E. An Automated Method for the Determination of Orthophosphate in the Presence of Labile Polyphosphates. *Soil Sci. Soc. Am. J.* **1980**, *44*, 173–175. [CrossRef]

28. Rydin, E. Potentially mobile phosphorus in Lake Erken sediment. *Water Res.* **2000**, *34*, 2037–2042. [CrossRef]

29. Zhou, B.; Vogt, R.D.; Lu, X.Q.; Xu, C.Y.; Zhu, L.; Shao, X.L.; Liu, H.L.; Xing, M.N. Relative Importance Analysis of a Refined Multi-parameter Phosphorus Index Employed in a Strongly Agriculturally Influenced Watershed. *Water Air Soil Pollut.* **2015a**, *226*, 25. [CrossRef]

30. Zheng, Y.; Zhou, L.Y.; Wang, X.; Shen, B.Z. Investigation and evaluation on some elements in the substrate sludge of Yuqiao Reservoir. *Acta Agric. Boreali-Sin.* **1991**, *6*, 122–126.

31. van Dael, T.; De Cooman, T.; Verbeeck, M.; Smolders, E. Sediment respiration contributes to phosphate release in lowland surface waters. *Water Res.* **2020**, *168*, 115168. [CrossRef] [PubMed]

32. Wang, N.L.; Wang, J.M.; Li, H.; Zhou, B.; Xing, M.N.; Liu, H.L. The Accumulation Characteristic of Sedimentary Phosphorus and Its Release Potential in Yuqiao Reservoir. *Environ. Prot. Sci.* **2020**, *46*, 56–61.

33. Smith, E.A.; Mayfield, C.I.; Wong, P.T.S. Physical and chemical characterization of selected natural apatites in synthetic and natural aqueous solutions. *Water Air Soil Pollut.* **1977**, *8*, 401–415.

34. Jaiswal, D.; Pandey, J. Hypoxia and associated feedbacks at sediment-water interface as an early warning signal of resilience shift in an anthropogenically impacted river. *Environ. Res.* **2019**, *178*, 108712. [CrossRef] [PubMed]

35. Yu, P.; Wang, J.; Chen, J.; Guo, J.; Yang, H.; Chen, Q. Successful control of phosphorus release from sediments using oxygen nano-bubble-modified minerals. *Sci. Total Environ.* **2019**, *663*, 654–661. [CrossRef] [PubMed]

36. Tammeorg, O.; Nürnberg, G.; Horppila, J.; Haldna, M.; Niemistö, J. Redox-related release of phosphorus from sediments in large and shallow Lake Peipsi: Evidence from sediment studies and long-term monitoring data. *J. Great Lakes Res.* **2020**, *46*, 1595–1603. [CrossRef]

37. Yang, C.; Yang, P.; Geng, J.; Yin, H.; Chen, K. Sediment internal nutrient loading in the most polluted area of a shallow eutrophic lake (Lake Chaohu, China) and its contribution to lake eutrophication. *Environ. Pollut.* **2020**, *262*, 114292. [CrossRef] [PubMed]

38. Wang, Y.M.; Li, K.F.; Liang, R.F.; Han, S.Q.; Li, Y. Distribution and Release Characteristics of Phosphorus in a Reservoir in Southwest China. *Int. J. Environ. Res. Public Health* **2019**, *16*, 303. [CrossRef] [PubMed]
39. Alam, M.S.; Barthod, B.; Li, J.Y.; Liu, H.; Zastepa, A.; Liu, X.C.; Maria, D. Geochemical controls on internal phosphorus loading in Lake of the Woods. *Chem. Geol.* **2020**, *558*, 119873. [CrossRef]

MDPI
St. Alban-Anlage 66
4052 Basel
Switzerland
Tel. +41 61 683 77 34
Fax +41 61 302 89 18
www.mdpi.com

Water Editorial Office
E-mail: water@mdpi.com
www.mdpi.com/journal/water